# The Ethics of Climate Governance

# The Ethics of Climate Governance

Edited by
Aaron Maltais and Catriona McKinnon

London • New York

Published by Rowman & Littlefield International, Ltd.
Unit A, Whitacre Mews, 26-34 Stannary Street, London SE11 4AB
www.rowmaninternational.com

Rowman & Littlefield International, Ltd. is an affiliate of Rowman & Littlefield
4501 Forbes Boulevard, Suite 200, Lanham, Maryland 20706, USA
With additional offices in Boulder, New York, Toronto (Canada), and London (UK)
www.rowman.com



**British Library Cataloguing in Publication Information Available**
A catalogue record for this book is available from the British Library

ISBN: HB 978-1-78348-214-6
ISBN: PB 978-1-78348-215-3

**Library of Congress Cataloging-in-Publication Data**

Library of Congress Cataloging-in-Publication Data Available
ISBN 978-1-78348-214-6 (cloth : alk. paper) -- ISBN 978-1-78348-215-3 (pbk. : alk. paper) -- ISBN 978-1-78348-216-0 (electronic)

∞™ The paper used in this publication meets the minimum requirements of American National Standard for Information Sciences Permanence of Paper for Printed Library Materials, ANSI/NISO Z39.48-1992.

Printed in the United States of America

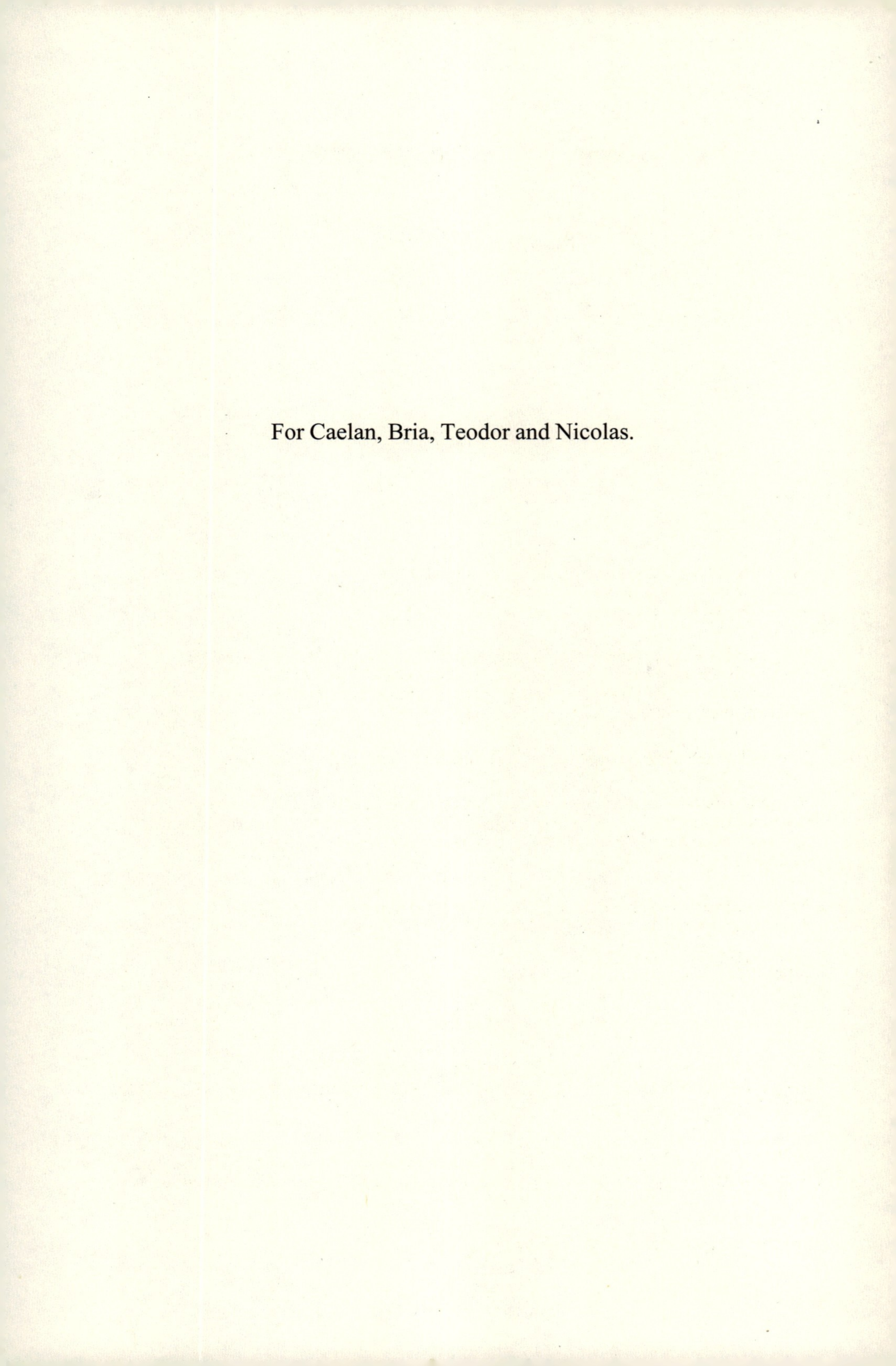

For Caelan, Bria, Teodor and Nicolas.

# Contents

# Introduction

## *The Ethics of Climate Governance*

Aaron Maltais and Catriona McKinnon

At the time of writing, nine months remains until the UNFCCC's COP 21 meeting in Paris at which parties aim to agree on a successor to the Kyoto Protocol. This should be an agreement with 'legal force' (UFCCC 2014: Decision 1/CP.20) that will 'bind nations together into an effective global effort to reduce emissions rapidly enough to chart humanity's longer-term path out of the danger zone of climate change, while building adaptation capacity'.[1] This is taking place against a background of an increasing consensus that without additional strenuous mitigation efforts there will be significant overshoot of the 2 degrees Celsius threshold to which nearly all of the world's states have committed themselves (e.g., UNFCCC 2010: Decision 1/CP.16). In the absence of mitigation it is projected that global mean surface temperature will increase by 3.7 to 4.8 degrees Celsius against pre-industrial levels by the end of this century (IPCC 2013). It is not hyperbolic to anticipate the winter 2015 meeting in Paris as the global community's last real chance to effect the mitigation goals it has set for itself. And it is not pessimistic to fear that the chance will be missed.

Broadly understood, 'climate governance' refers to the set of institutions, agreements, processes, practices and structures that enable the mitigation of, and adaptation to, climate change. This can be pursued by means of political action and agreement, legal measures, economic restructuring and market reform, administrative and bureaucratic evolution, policy initiatives, social reform, and the sharing of information. The scale of these activities can be fully global (e.g., the Kyoto Protocol), transnational and regional (e.g., the EU Emissions Trading Scheme), domestic (e.g., the UK Climate Change Act 2008), metropolitan and urban (e.g., the C40 Cities Climate Leadership

Group),[2] or very small scale and community based. Finally, the agents of climate governance are similarly diverse: global institutions (e.g., the UNFCCC), regional institutions (e.g., the Climate for Development in Africa group, ClimDev-Africa), states, cities, activist groups, and even school networks.

Given the urgency and the severity of climate change as a 'super wicked' problem (Levin et al. 2012) it is tempting to take an entirely instrumentalist approach to thinking about climate governance by asking only what forms and institutions of governance will be effective, at the lowest cost, in the shortest possible time frame.[3] But a moment's thought shows that questions of effectiveness, costs and time frames require ethically informed reflection. Different types of climate governance have different beneficial effects, which may be unevenly distributed across the globe and through time, and which have different economic, political, social and cultural costs. The question of how these benefits and costs ought to be distributed and allocated across present and future people who will be affected by climate change is an ethical question. Much good work has been done in this area in recent years, to the extent that there is now an identifiable field called 'climate ethics' that did not exist a decade ago (e.g., Gardiner et al. 2010).

Work on the ethics of climate change has been particularly focused on assessing principles for how the costs of mitigation, adaptation, and compensations should be distributed between wealthier and poorer regions and between present and future generations. What this work makes clear is that we are very far from achieving climate justice. Humanity is currently on track to increase global greenhouse gas (GHG) emissions by fifty percent over current levels by 2040 (Marchal et al. 2012, 72). A realistic chance of meeting the 2 degree target requires that we cut emissions levels by 50 percent from current levels by 2050 (IPCC 2013, 12). Because this reality threatens to undermine our ability to avoid serious climate damages and to support all regions in adapting, it has become increasingly clear that work on the ethics of climate change must start to pay much more attention to questions of *how* to actually put effective and equitable climate governance into practice (e.g., Shue 2011, 2013; Stevenson & Dryzek 2012; Eckersley 2012; Maltais 2014; Gardiner 2014; Caney 2015). This book, based on a collection of papers, addresses that topic: when thinking about how to design instruments of climate governance, what ethical considerations ought to constrain and direct proposals, and what difference do these considerations make to the cogs and wheels of these instruments? The diversity and scope of climate governance initiatives and issues put a comprehensive analysis of all of them well beyond the scope of a single book. This book collects together pairs of chapterss on four core topics in the ethics of climate governance. These topics are: the problem of vulnerability and domination in international climate governance; potential tensions between democratic values and effective climate governance; the challenges of motivating present generations to invest in avoiding future climate impacts; and the ethical and governance dimensions of new technologies expected to play large roles in addressing the threat of global warming.

Part I, 'Domination and Vulnerability in International Climate Governance', evaluates the international dimensions of climate governance given a commitment to protecting vulnerable present and future people from domination and exploitation. Megan Blomfield explores a neglected dimension of present (and future) inequities in the distribution of natural resources (inter alia, emissions allocations) by considering them as the product of the unjust governance of natural resources in the past. Taking colonialism as her example, she argues that some of these non-rectified past injustices have very likely left some present groups particularly vulnerable to the impacts of climate change, and are badly placed to shape climate agreements—as they are presently negotiated—to address their vulnerabilities. The question of how a global climate treaty risks exploiting the vulnerabilities of such groups is taken up by Patrick Taylor Smith. He argues that protecting vulnerable groups from the worst effects of climate change through effective mitigation and adaptation sets constraints on any global climate regime. Such a regime would have to have significant legislative, judicial and executive authority. But any such regime would itself be at serious risk of becoming dominating. Taylor Smith explores strategies to minimise this risk by making a global climate regime accountable, and the potential costs of these strategies with respect to the demands they make of individuals, and the effectiveness of the treaties the regime enables.

In Part II the focus shifts from vulnerability and domination to democracy and legitimacy. Democratic institutions have well-documented virtues: at their best they realise equality, enable participation, enhance voice and promote freedom. However, in the face of the super wicked problem of climate change, and given the likelihood that some form of global governance will be necessary for mitigation and adaptation (even if this does not exhaust all the available options for climate governance), we might start to doubt that democratic values should take priority in the principles of climate governance. Ludvig Beckman addresses a specific example of this line of thought that focuses on human rights, as follows. In governing climate change, it is of primary importance to ensure that human rights are protected. If institutions that protect human rights are impeded by institutions that realise democratic procedures, then the latter ought to give way. Beckman argues that this line of thought is mistaken and that duties to uphold human rights in the face of climate change, and duties to uphold democratic institutions, are not in conflict with respect to what is owed to present people by present people. With respect to future people, the picture is more complicated, and the duties of present people to uphold future people's future human rights may indeed be in tension with present people's duties to uphold democratic institutions. This tension created by the moral relationships between present and future people is one of the features of climate change as a moral problem that makes it so difficult to govern (and the theme is taken up in Maltais's chapter in the next section).

Another significant set of doubts about using decision-making institutions and processes in the governance of global climate change is a product of the urgency of the problem, combined with the diversity and number of interests feeding into the process. In seeking agreement on mitigation and adaptation proposals, must we choose between inclusive agreement on minimally demanding proposals, on the one hand, and more limited agreement on more demanding proposals, on the other hand? If democratic deliberation about a solution to a problem requires including in the deliberation as many parties as possible affected by that problem, then we might think that such deliberation ought to be abandoned, or at least postponed, in the face of a super wicked problem like climate change. The reasoning here is that including as many affected parties as possible will at best deliver agreement only on shallow proposals, if we take interests as fixed. Given the time frame in 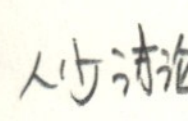which an agreement must be found, it might be thought that moving to a less inclusive 'minilateral' or 'club' model for agreement is preferable. Here, meaningful and effective proposals are possible because of how participation in negotiation processes is limited to a smaller number of participants whose interests are more likely to be aligned. In his chapter, Jonathan Kuyper argues that rather than being an impediment to action, inclusivity in climate negotiations is a prerequisite for the effectiveness of the outcome of the process, independent of whether inclusivity is also required by justice. He makes this argument by appeal to two benefits of inclusive multilateral decision-making: that it produces epistemically superior outcomes, and that it enhances compliance. If Kuyper is right, then the contrast between minilateral negotiations producing deep and effective outcomes, and multilateral negotiations producing shallow and ineffective outcomes, is false.

In Part III the problem of motivating individuals and governments to support and take actions to significantly reduce GHG emissions is addressed. Almost all ethically informed analyses of the problems created by climate change, and the costs and benefits of proposed solutions, acknowledge that although all existing people will be affected by climate change it is young and future people that are most at risk from the consequences of global warming. There are difficult questions about how to ensure that the interests of people yet to come into existence are properly represented in climate negotiations between present people. If these negotiations were to issue in a global action plan enabling conformity with the IPCC's most optimistic scenarios—wherein emissions drop off steeply, a zero carbon economy is achieved by 2100, and adaptation is co-ordinated, effective and well resourced—then future people would still suffer the worst impacts of climate change, but not in a catastrophic or very severe way (IPCC 2013: 21). However, optimism may not be warranted given the scale of the political and economic changes (and the inevitable consequences for the lifestyles of the world's most affluent nations and people) required to shift on to such a trajectory. There appears to be a real risk that present people not facing the worst effects of climate change will not be willing to bear the costs of mitigation required to limit warming to around 2 degrees Celsius. Taking seriously the prospect that there are strong conflicts of interests between present and future generations, Aaron Maltais considers arguments for permitting present people to achieve mitigation by 'borrowing from the future'. Although he is sympathetic to not ruling out this line of thought from the portfolio of ethically acceptable approaches to climate governance, he is sceptical about whether debt financing proposals would indeed be fit to motivate present people to mitigate climate change. If this is not the case, Maltais argues that we may need to consider alternative 'pre-commitment' strategies that bind the young of the present, and near subsequent, generations to financing mitigation as a form of insurance against a pattern of perpetually delaying meaningful mitigation investments.

Decarbonisation of the global economy is now beyond serious dispute as a necessary part of meaningful mitigation. In his chapter, Steve Vanderheiden argues 'informational governance'—that is, programmes to gather, reveal and disseminate information about the emissions footprints of various activities, especially those undertaken by businesses—is a key measure for achieving decarbonisation. He argues that informational governance has benefits that can complement regulatory mechanisms to limit emissions by enhancing political accountability and personal responsibility. In particular, he focuses on how informational governance that gives people knowledge of the social and environmental effects of their consumption and lifestyle choices can generate transformation in their motivation and practice, in the name of maintaining personal integrity. None of this is straightforward or easy to achieve, and Vanderheiden considers different types of informational governance by comparing their benefits and drawbacks. But ultimately, unless the global affluent in particular are motivated to drastically reduce their emissions footprints (and to pressurise their governments to make it easier for them to do so), decarbonisation will be impossible. Informational governance could be one important way in which to grow these motivational states.

Part IV of the book turns to new technologies that many think will become a large part of mitigation and adaptation solutions by the end of this century. Perhaps the most well-known set of new technologies for mitigation and adaptation are in the area of geoengineering. Broadly speaking, geoengineering refers to the large scale manipulation of the Earth's planetary systems, with a deliberate purpose. Geoengineering techniques range from familiar practices such as afforestation (which, if done on a massive scale, could significantly reduce emissions by creating carbon sinks) to seemingly sci-fi proposals that involve putting mirrors into space to deflect solar radiation. All geoengineering proposals raise governance problems albeit differently cast according to the technology in question. For example, solar radiation management (SRM) techniques such as injecting aerosols (a common suggestion is sulphur particles) into the stratosphere could conceivably be undertaken by a state acting unilaterally, whereas carbon capture techniques could only ever be effective in mitigation if their use was widespread across all states. In the first case, a key feature of the problem is how the community of states could coordinate to regulate action affecting them all. In the second case, the global governance problem involves coordination and, arguably, an equitable distribution of the costs and benefits of the techniques.

In her chapter, Clare Heyward asks whether the debate about geoengineering—in particular, sulphate aerosol injection—conforms to 'the novelty trap' evident in past debates about new technologies. Steve Rayner (2004) adopted the term 'novelty trap' to describe a dynamic where proponents of a new technology make extravagant claims about its transformative impact, skeptics claim that the far reaching impact of this new technology brings with it extraordinary risks, and proponents respond to charges of novel risk by claiming continuity in this new technology with past technology and experience. Heyward argues that when discussion about a new technology has this character, progress towards discussion of ethically urgent questions about its implementation is impeded; avoiding the red herrings of the novelty trap is necessary for needed and productive discussion about the governance of geoengineering.

The IPCC includes bioenergy with carbon capture and storage (BECCS) in its Representative Concentration Pathway (RCP) 2.6 scenario, in which warming is minimised. The main advantage of BECCS is that it produces negative $CO_2$ emissions without decreasing energy output. Although the large-scale infrastructure to enable BECCS as a central part of global mitigation efforts is not in place, or even in train, the technology already exists. The bioenergy production central to BECCS is commonly achieved through biomass incineration. In her contribution to the collection, Kristin Shrader-Frechette focuses on the biofuel Miscanthus giganteus, which is one of the most prominent energy crops in the EU, and which holds most promise for the production of biomass necessary for BECCS. Shrader-Frechette raises serious concerns about the ethical and health-related hazards of Miscanthus, and examines the consequences of these concerns for democratically acceptable decisions about where and how the crop is grown and processed. Her aim is to emphasise that our existing noncarbon energy technologies—which already have relatively well-developed infrastructures—remain the most promising and acceptable way forward for decarbonisation.

To sum up, all of the chapters in this collection mark out territory that will be a permanent feature of climate governance for decades to come, if not longer. What unites them and makes them distinctive is the central place they give to ethical considerations relevant to the principles and choice of mechanisms for climate governance. We know from history that the outcomes of political processes in large part inherit their moral qualities from those possessed by the processes themselves. In this respect, climate governance is continuous with all other forms of governance. What marks the difference in the case of climate change is what is at stake for humanity as a whole, now and in the future, if we miss present opportunities to govern our collective responses well.

## REFERENCES

Caney, Simon. 2015. 'Responding to Noncompliance'. In *Climate Justice in a Non-Ideal World*, eds. Dominic Roser and Clare Heyward. Oxford: Oxford University Press.

Eckersley, Robyn. 2012. 'Moving Forward in Climate Negotiations: Multilateralism or Minilateralism'? *Global Environmental Politics* 12: 24–42.

Gardiner, Stephen M. 2014. 'A Call for a Global Constitutional Convention Focused on Future Generations'. *Ethics & International Affairs*, 28: 299–315.

Gardiner, Stephen M., Simon Caney, Dale Jamieson, and Henry Shue, eds. 2010. *Climate Ethics: Essential Readings*. Oxford: Oxford University Press.

IPCC. 2013. Summary for Policymakers. In *Climate Change 2013: The Physical Science Basis. Contribution of Working Group I to the Fifth Assessment Report of the Intergovernmental Panel on Climate Change* [Stocker, T.F., D. Qin, G.-K. Plattner, M. Tignor, S.K. Allen, J. Boschung, A. Nauels, Y. Xia, V. Bex and P.M. Midgley (eds.)]. Cambridge University Press, Cambridge, United Kingdom, and New York.

Levin, Kelly, Benjamin Cashore, Steven Bernstein, and Graeme Auld. 2012. 'Overcoming the Tragedy of Super Wicked Problems: Constraining Our Future Selves to Ameliorate Global Climate Change'. *Policy Sciences* 45: 123–52.

Maltais, Aaron. 2014. 'Failing International Climate Politics and the Fairness of Going First', *Political Studies*. 62: 618–33.

Rayner, Steve. 2004. 'The Novelty Trap: Why Does Institutional Learning about New Technologies Seem So Difficult?' *Industry and Higher Education* 18: 349–55.

Shue, Henry. 2011. 'Face Reality? After You!—A Call for Leadership on Climate Change'. *Ethics and International Affairs* 25: 17–26.

———. 2013. 'Climate Hope: Implementing the Exit Strategy'. *Chi. J. Intl Law* 13: 381–401.

Stevenson, Hayley, and John S. Dryzek. 2012. 'The Legitimacy of Multilateral Climate Governance: A Deliberative Democratic Approach'. *Critical Policy Studies* 6: 1–18.

UNFCCC. 2010. 'Report of the Conference of the Parties on Its Sixteenth Session. Part Two: Action Taken by the Conference of the Parties at Its Sixteenth Session'. http://unfccc.int/resource/docs/2010/cop16/eng/07a01.pdf (accessed 20 February 2015).

UNFCCC. 2014. 'Report of the Conference of the Parties on Its Twentieth Session. Part Two: Action Taken by the Conference of the Parties at Its Twentieth Session'. http://unfccc.int/resource/docs/2014/cop20/eng/10a01.pdf (accessed 20 February 2015).

## NOTES

1. United Nations Framework Convention on Climate Change, 'Warsaw Outcomes', accessed 19 February 2015, http://unfccc.int/key_steps/warsaw_outcomes/items/8006.php.

2. http://www.c40.org/

3. 'Super wicked' social problems are those requiring a solution when time is running out, when there is no central authority that can coordinate the solution, where those who have to solve the problem are also those who are causing it, and where discounting of the future creates incentives to postpone responses that could solve the problem.

# *I*

# Domination and Vulnerability in International Climate Governance

## *Chapter One*

# Climate Change and the Moral Significance of Historical Injustice in Natural Resource Governance[1]

Megan Blomfield

In discussions about responsibility for climate change, it is often suggested that the historical use of natural resources is in some way relevant to our current attempts to address this problem fairly. In particular, both theorists and actors in the public realm have argued that historical high emitters of greenhouse gases (GHGs)—or the beneficiaries of those emissions—are in possession of some form of debt, deriving from their overuse of a natural resource that should have been shared more equitably. These accounts of what might be termed 'natural debt' generally focus on one particular natural resource (global GHG sink capacity); invoke a principle of justice by which rights to consume this resource should have been allocated (most commonly, equal per capita shares); and then argue that historical violations of this principle give rise to certain rectificatory duties in the present (generally, duties on the part of those who have historically consumed an excessive amount of the world's GHG sink capacity, or who have benefitted from such excess consumption, to offer some form of compensation to those who have not—such compensation usually taking the form of emission credits or cash).[2]

Though many seem to find it intuitively plausible that historical high emissions have incurred some form of debt, significant challenges arise in rendering the concept of natural debt both coherent and defensible.[3] Such problems will not, however, be my focus in this chapter. Instead, I here suggest that discussions about historical responsibility for climate change commonly fail to recognise certain other past injustices concerning natural resources that appear to hold contemporary relevance. In particular, I argue that it is not just the unequal consumption of global GHG sink capacity that may be of moral significance here, but also the way in which the world's resources have more generally been governed.

In order to address the matter of historical responsibility for climate change, it is first important to be clear about the nature of the problem that climate change presents. It is this task that I take up in the first section of the chapter. Highlighting the issue of unequal risk, I explain how climate change *as a problem of global justice* has both physical and social contributors. Accounts of natural debt attempt to determine responsibility for the physical component of the problem, overlooking the question of who should be held responsible for the inequalities that make climate change such a challenging problem of global justice.

As I explain in the second part of the chapter, this latter question is more difficult to answer, because a plethora of historical injustices appear to be causally implicated in the fact that some communities are significantly more vulnerable to climate events than others. And in any case, even where we can conclude that certain historical injustices have causally contributed to the problem of climate change, using this conclusion to determine who should bear the costs of climate action remains far from straightforward. When current situations ought to be rectified due to historical injustice, the demand for rectification holds independently of whether or not those situations are now contributing to the problem of climate change. And if socio-economic circumstances place communities at risk of disaster or undermine our attempts to deal with climate change fairly, this problem must be dealt with regardless of the historical provenance of those circumstances.

In the final sections of this chapter I argue that attention to historical injustice is nevertheless necessary if we are to deal with climate change fairly. Using colonialism as an example, I argue that the causal links between climate vulnerability and historical wrongs suggest that in some respects, the problem of climate change is actually part of an on-going, or enduring, injustice. It is only when an injustice is on-going in this way that we are confronted with the question of *why* it endures;[4] and when this question arises, we are given reason to search for persisting structures—whether practices, institutions, processes, systems or rules—that perpetuate injustices and prevent their rectification. In the case of climate change, I suggest, enduring systems of unjust natural resource governance appear to be one of the reasons why many historical wrongs of this kind have a continuing legacy of injustice in the present. If we do not address these problematic structures of resource governance, then there is good reason to fear that even well intentioned efforts to find a just solution to the problem of climate change will be undermined. Historical injustices of natural resource governance thus have contemporary moral significance because they are part of an enduring injustice that persists to the present day, and which now threatens to prevent us from dealing with climate change fairly.

## CLIMATE CHANGE AS A PROBLEM OF GLOBAL JUSTICE

Though climate change is widely understood to be a problem of global justice, it is not always made precisely clear why this is. Climate change can easily be recognised as an ethical problem, because it is an anthropogenic phenomenon that threatens significant harm to human beings. From this one can derive a moral obligation to reduce the risk of such harm by doing something to address climate change. The question of how to discharge this obligation then raises questions of both intergenerational and global justice: specifically, how to distribute the burdens of dealing with climate change between different generations and—within any given generation—between contemporaries who vary greatly in their capabilities and responsibilities.[5]

However, duties to address climate change can be derived not only from an imperative to protect (or refrain from harming) human beings, but also on grounds of global justice. As many have observed, climate change not only threatens important human interests, but also appears to do so in a way that is distinctly unfair. This is because climate change does not pose the same level of risk to all human beings. It is the global poor, both now and in the future, who tend to be most at risk from climate change—including poor communities that are located within affluent countries (IPCC 2007b, 19). Poor communities are especially at risk due to 'limited adaptive capacities' and greater dependence on 'climate-sensitive resources such as local water and food supplies'. Africa in particular is 'one of the most vulnerable continents'; and in Latin America the effectiveness of adaptation efforts is being 'outweighed' by the lack of 'appropriate political, institutional and technological frameworks [and] low income' (IPCC 2007a, 12–14).

Duties to address climate change can therefore also be understood as duties of global justice: as duties to prevent the impacts of climate events from causing our global circumstances to migrate ever further away from what justice demands. Proponents of the idea of natural debt and the closely related Beneficiary Pays Principle (BPP) then highlight what appears to be an added complexity to the ethical problem of climate change.[6] This is that, generally speaking, the world's poor are not only most at risk of harm, but in addition are the least responsible for—and have benefitted the least from—the GHG emissions that cause climate change. The world's rich, conversely, tend not only to be the least at risk, but also to be most responsible for—and the most benefitted by—GHG emissions. A number of theorists have suggested that this creates a further layer of injustice to climate change, because the benefits and burdens of a particular activity—namely, the emission of

GHGs—are being distributed very unevenly;[7] and because the countries most at risk of harmful climate impacts are the least responsible for contributing, causally, to the physical phenomenon of climate change.

One might also, however, look more closely at *why* it is that climate change places some human beings at greater risk of disaster than others.[8] As the IPCC's Special Report on Extreme Events explains, disaster risk can be better understood once we recognise that it emerges not from the threat of climate events alone, but from 'the interaction of weather or climate events, the physical contributors to disaster risk, with exposure and vulnerability, the contributors to risk from the human side' (IPCC 2012, ix). The first human determinant of disaster risk—exposure—describes the extent to which a given element (e.g., a population, its livelihood or other assets) is present in a place that could be adversely affected by climate events. The second—vulnerability—'refers to the propensity of exposed elements… to suffer adverse effects when impacted by hazard events'. Exposure is a *necessary determinant* of disaster risk; if a population or system is not exposed to climate events, then it is not vulnerable to, or at risk of, climate disaster. Exposure is not, however, a *sufficient* determinant of disaster risk. This is because it is possible to be exposed to climate events but not be vulnerable—for example, by being located in a flood plain but having the capacity to employ defences that will prevent a climate change-induced flood from creating significant losses (IPCC 2012, 69).

Vulnerability is thus a very important driver of disaster risk, but one that is commonly overlooked in discussions about climate change and historical responsibility. When theorists focus only on determining who can be taken to have emitted (or benefitted from) more than their fair share of GHGs, they effectively restrict their concern to the question of who should be held responsible for the *physical* determinant of disaster risk (the physical determinant, recall, being the climate events to which GHG emissions have causally contributed). But as the IPCC explains, 'climate change is not a risk per se'. Rather, the risk to which communities are subject arises from the interaction of climate changes with vulnerability and exposure (IPCC 2014a, 1050). So why not also consider historical responsibility for these necessary *social* determinants of disaster risk—and in particular, for vulnerability?

Perhaps one of the reasons that some theorists do not direct their attention to the question of who should be held responsible for the vulnerability component of disaster risk is that this propensity to suffer adverse effects derives from a very complex set of factors. Whilst the anthropogenic drivers of climate events are quite easy to identify—namely, human activities, like deforestation and the burning of fossil fuels, which lead to increased atmospheric concentrations of GHGs—vulnerability is 'a result of diverse historical, social, economic, political, cultural, institutional, natural resource, and environmental conditions and processes' (IPCC 2012, 32). A number of factors can, however, be singled out as major contributors to vulnerability. High vulnerability, the IPCC suggests, is 'mainly an outcome of skewed development processes, including those associated with environmental mismanagement, demographic changes, rapid and unplanned urbanization, and the scarcity of livelihood options for the poor'. Other contributing factors identified are 'poverty, and the lack of social networks and social support mechanisms'; and 'global processes' such as 'international financial pressures, increases in socioeconomic inequalities, trends and failures in governance (e.g., corruption, mismanagement), and environmental degradation' (IPCC 2012, 70-71). Vulnerability can also result from land tenure arrangements that create insecurity, or that leave certain groups marginalised (IPCC 2012, 306; see also IPCC 2014a, 1051).

It will be impossible to explain precisely why any given human community is beset by factors that render it more or less vulnerable in the face of climate events; or—therefore—to determine exactly who should be held responsible for the fact that climate change poses much greater risk to some communities than others. Nonetheless, in the next section I argue that any adequate explanation of the social inequalities that engender the uneven distribution of vulnerability to climate change must acknowledge that numerous historical injustices appear to have played a role in creating this state of affairs. I then discuss how such historical injustice could hold moral significance for our current attempts to govern the climate change problem fairly.

## CLIMATE CHANGE AND HISTORICAL INJUSTICE

In order to figure out who should be held responsible for the fact that some human populations are particularly at risk from climate change, it seems that we must explain why it is that some communities are burdened by problems of underdevelopment; environmental degradation and mismanagement; poverty, inequality and scarcity of livelihood options; institutional weakness and failed governance; vulnerability to international financial pressures; or land tenure arrangements that engender insecurity and marginalisation. Such explanations will be difficult to provide and will necessarily differ for each community. But one thing of which we can be sure is that various historical injustices will be causally implicated in the fact that certain communities are afflicted by factors that render them particularly vulnerable to climate events.

As Thomas Nagel famously says, 'We do not live in a just world . . . may be the least controversial claim one could make in political theory' (2005, 113). Such injustice becomes even more apparent when we look back through history to consider how wrongs including unjust war, colonialism and slavery have helped to bring us to our current state of affairs. Each of these forms of injustice will have played a role in rendering some communities particularly vulnerable to climate events; for example, through legacies of environmental degradation, failed governance or poverty and inequality. In what follows, however, I will be focussed on the example of colonialism due to the extremely significant role that colonialism played in the history of global resource governance.

The suggestion that the legacies of colonialism could be relevant to our current attempts to deal with climate change fairly is not a novel one. Henry Shue long ago suggested that 'colonial exploitation' could have contributed to what he terms the 'background injustice' of the climate negotiations. As Shue notes, poverty can make countries vulnerable both to climate events and to manipulation in the climate negotiations. Many poor countries do not have the resources to cope with unmitigated climate change, leaving them in such desperate need of *some* international agreement on climate action that they 'might have no better alternative' than to concede to 'unconscionable terms' (Shue 1992, 387–88). Insofar as colonial exploitation has engendered such poverty—and, thus, climate vulnerability—it has therefore also helped to place certain parties to the UNFCCC at an unfair disadvantage in the climate negotiations. Stephen Gardiner—referencing Shue—similarly suggests that 'the history of colonialism' has contributed to the serious injustice of the 'existing world system', injustice that will undermine fair climate governance by enabling 'powerful countries' to take 'further advantage of those already exploited under the current structure' (Gardiner 2011, 119).[9]

Another theorist who suggests that colonialism has contributed to the problem of climate change is Robert Melchior Figueroa. Focussing on indigenous peoples in particular, Figueroa argues that factors including 'colonial practices of resource exploitation, relocation, land appropriation [and] persistent economic exploitation'—and their ongoing legacies of struggles for self-determination, under-representation in environmental decision-making and distributive inequities—'capture the causal roots of precisely why indigenous groups are the most vulnerable and impacted by climate change' (2011, 235–36). Some support for Figueroa's claim can be found in the latest IPCC report, where it is noted that indigenous peoples in North America are vulnerable in part because 'the legacy of their colonial history . . . has stripped Indigenous communities of land and many sources of social and human capital' (IPCC 2014b, 1471).

Despite these suggestions, the causal contribution of colonialism to the problem of climate change—and the present moral significance of such contribution—remains underexplored in the philosophical literature on climate justice and responsibility. Shue provides some explanation of this lacuna when he offers his own reasons for not pursuing the matter further. Though the vulnerability of some parties to the climate negotiations is clearly enhanced by their poverty, the extent to which such poverty results from colonialism depends on what Shue takes to be 'important but intractable debates about causal mechanisms' (1992, 391). The causal mechanism in the case of GHG emissions, on the other hand, is much harder to dispute: historical contributions to increased atmospheric concentration of GHGs—and, therefore, to the physical phenomenon of climate change—can be identified and quantified relatively easily. Plausibly, this is the reason why most accounts of historical responsibility for the problem of climate change have focussed on the contribution of past emissions rather than historical injustices.

Nevertheless, as Shue also points out, 'causal responsibility does not translate smoothly into moral responsibility' (1992, 391, fn. 11). It could therefore be the case, as far as climate change is concerned, that though the case for *causal* responsibility may be made more easily with respect to past emissions, a case for some form of *moral* responsibility could be on stronger ground in instances where it can be shown that clear historical wrongs are causally implicated in present vulnerability. In what follows, I start by attempting to show that colonialism plausibly *is* causally implicated in many cases of absolute and relative vulnerability to climate change. I then discuss the moral significance of such causal responsibility.

## HOW COLONIALISM IS CAUSALLY IMPLICATED IN CLIMATE VULNERABILITY

As Daniel Butt states, it is 'a truism to say that we live in a world that has been deeply shaped by imperialism. The history of humanity is, in many ways, a story of the attempted and achieved subjugation of one people by another, and it is unsurprising that such interaction has had profound effects on the contemporary world' (2012, 227–28). In this section, I draw on the work of Daron Acemoglu, Simon Johnson and James Robinson to provide reasons to think that colonial subjugation in particular has made a significant contribution to current levels of vulnerability to climate change.

Historically, colonialism took a variety of forms. In some cases the indigenous population was exterminated completely, in others it was displaced and dispossessed, or exploited as a source of labour. In North America, Australia and New Zealand, colonisers settled in the territory and set up political institutions resembling those of Europe; in other places, very little settlement occurred and colonies were essentially exploited as a source of income, wealth, and natural resources. This latter form of economic colonialism took place across 'much of Africa, Central America, the Caribbean, and South Asia' (Acemoglu 2003, 27), where colonial powers set up or took over what Acemoglu, Johnson and Robinson term 'extractive institutions'. Extractive institutions are 'bad and dysfunctional institutions' designed to support 'the extraction of resources by one group at the expense of the rest of society' (Acemoglu, Johnson and Robinson 2006, 21). In extractive colonial states, the main purpose of such institutions was of course 'to transfer as much of the resources of the colony to the colonizer' as possible (Acemoglu, Johnson and Robinson 2001, 1370); the 'resources' in question including precious minerals such as gold and diamonds; the products of plantation agriculture; human beings; and other sources of income and wealth.

There are a number of reasons to think that historical practices of colonialism continue to have pernicious effects in the present day, long after the UN's 1960 Declaration on the Granting of Independence to Colonial Countries and Peoples. Powers of self-determination, once destroyed or obstructed, are not easily realised even when formally protected by international law, and various colonial policies will have made the exercise of this collective capacity even more difficult. The institutions established by colonial authorities are likely to possess elements of path-dependence that make it hard for previously colonised peoples to alter their developmental trajectory. Thus, even long after independence, the choices faced by such collectives may be 'constrained by decisions that colonial masters made on their behalf' (Ypi, Goodin and C. Barry 2009, 127–29, 132).

Acemoglu, Johnson and Robinson suggest that path-dependence may have been particularly strong in countries where colonisers set up or took over extractive institutions. In general, very few constraints were placed on political power in these extractive colonies. Rather, colonisers intentionally created 'authoritarian and absolutist states with the purpose of solidifying their control and facilitating the extraction of resources' (Acemoglu, Johnson and Robinson 2001, 1375), placing 'a high concentration of political power in the hands of a few who extracted resources from the rest of the population' (Acemoglu, Johnson and Robinson 2002, 1264). Such institutions 'have a lot of staying power' because even after independence, the elites of extractive societies—'who benefit from using the power of the state to expropriate others'—will have much to lose from reform, and are therefore likely to 'resist and attempt to block any move toward better institutions'. As a result, many extractive institutions originally established in colonial times persist into the present day and continue to have adverse economic effects on the countries in which they are located (Acemoglu, Johnson and Robinson 2006, 31). Quoting Crawford Young, Acemoglu, Johnson and Robinson conclude that the supposedly 'new states' that emerged from independence were often really 'successors to the colonial regime, inheriting its structures, its quotidian routines and practices, and its more hidden normative theories of governance' (2006, 31; quoting C. Young 1994, 283).

Interestingly in the context of climate change, Acemoglu, Johnson and Robinson claim that the most significant long-term economic effects of extractive colonial institutions are a result of the role that they played during the Industrial Revolution. Elites in extractive states had reason to block industrialisation due to fears that it would undermine their position of power by benefiting and strengthening entrepreneurial and skilled members of the non-elite, and giving rise to political disruption (Acemoglu, Johnson and Robinson 2002, 1273). So while colonial authorities 'sowed the seeds of underdevelopment in many diverse corners of the world by imposing, or further strengthening existing, extractive institutions' (Acemoglu and Robinson 2012, 250); they also—through those same institutions—created current global inequality by ensuring that 'during the nineteenth and twentieth centuries some nations were able to take advantage of the Industrial Revolution while others were unable to do so' (Acemoglu and Robinson 2012, 271). Colonialism thus 'not only explains why industrialization passed by large parts of the world but also encapsulates how economic development may sometimes feed on, and even create, the underdevelopment in some other part of the domestic or the world economy' (Acemoglu and Robinson 2012, 273).

The material effects of colonial injustice would have been difficult to counter after independence since, as Butt points out, 'it is hard to acquire alternative entitlements once one has been unjustly deprived of large quantities of one's natural resources and/or is at a competitive trading disadvantage relative to other nations' (2009a, 113). Such disadvantage at the international level seems likely to have been further reinforced by the global order that emerged from imperialism, an order dominated by rich and powerful agents, where unjust inequalities could be sustained and enhanced. As Mathias Risse points out, the current 'global political and economic order' . . . 'emerged from the spread of European control since the fifteenth century'; where 'even systems that escaped Western imperialism had to follow legal and diplomatic practices imposed by Europeans' (2005, 9). Today, this global order is shaped by 'economically powerful states' and institutions like the World Bank, IMF, and WTO (Risse 2005, 9); and the same Western powers that were responsible for colonialism have been able to mould international law 'so as to secure and legitimate their own advantages—advantages which were often improperly obtained' (Butt 2009b, 163–64).

In the face of widespread historical injustice, 'the relative prominence and bargaining power of precisely those countries most responsible for the commission of said injustice in the development of international law' (Butt 2009b, 171)—and in the development of the world economy—suggests that such injustice has significant lasting effects; that the global order continues to be governed by structures that were 'developed on the terms of the affluent states, and shaped in their interests' (Butt 2009b, 171). All of these factors may help to explain why it remains the case that 'some of the poorest countries in the world are former colonies of some of the richest' (Butt 2012, 230).

The above hopefully makes clear that there are a number of ways in which colonialism will be causally implicated in the current vulnerability of some human communities to climate change. This historical injustice plausibly has continuing legacies that include self-determination struggles; institutional weakness, authoritarian governance and elite capture; persisting land-tenure arrangements that breed marginalisation and insecurity; underdevelopment and lack of industrialisation; enduring problems of poverty, social inequality and lack of diversity in livelihood options; environmental degradation and mismanagement resulting from the overexploitation of land and natural resources; and vulnerability to global economic and political pressures in an international order structured by rich and powerful (and often ex-imperial) states. Against this historical background, climate change can be seen as a phenomenon that is exacerbating pre-existing, unrectified injustices, many of which derive from the same historical process that created such unequal rates of industrialisation—and, thus, GHG emissions—in the first place.

Drawing on empirical data, J. T. Roberts and Bradley Parks reach a similar conclusion. Referring in particular to cases where imperial powers structured colonial economies around the 'extraction of raw materials',[10] they argue that 'many of the most important causal forces driving hydrometeorological risk—from declining terms of trade and deteriorating infrastructure to degraded natural environments and weak and corrupted political institutions—are a *direct consequence* of extractive colonial legacies' (2007, 104–105; emphasis added).[11] Jon Barnett and John Campbell likewise conclude that colonialism served to increase vulnerability in Pacific Island communities by reducing agricultural diversity, introducing less resistant crops, and replacing land that was used for local food production with plantations and commercial agriculture. In this case, furthermore, colonial authorities (and missionaries) also helped to increase climate *exposure*, by encouraging communities—traditionally situated inland, on higher ground—to move to coastal regions, and establishing urban administrative centres on the coast (Barnett and Campbell 2010, 34–35). And Emilie Cameron argues that it is a failing for research that 'identifies rapid social, cultural, political, and economic change over the past decades as an important component of Inuit vulnerability to climatic change', not to 'explicitly name these changes as tied to colonialism' (2012, 109).

Thus, it would appear that whilst historical GHG emissions are indeed relevant to climate change (being, as they are, a major contributor to the physical phenomenon), colonialism is also causally implicated in this problem. Colonialism has helped to create a world characterised by severe inequalities in vulnerability to climate events (and, thus, disaster risk), and in doing so it has also helped to make climate change a particularly challenging problem of global justice. In cases where the risks to which colonial practices have contributed ripen into actual disasters, both historical emissions and colonial practices will have some share of causal responsibility for the resulting harms.

## THE MORAL SIGNIFICANCE OF COLONIALISM'S CONTRIBUTION TO THE PROBLEM

The picture of causal responsibility for climate disasters that emerges from the previous section is a complicated one. We find that when communities are subject to climate impacts, causal responsibility for this harm will be shared not only between those who engaged in activities (such as deforestation and the burning of fossil fuels) that enhanced atmospheric concentrations of GHGs, but also those who contributed to that community's vulnerability and exposure to climate events. This latter category will include both domestic and international agents and, in many cases (as I argued in the previous section), former colonial powers.

The *moral* significance of such *causal* responsibility nevertheless remains to be determined. It is certainly important that the link between colonialism and vulnerability is acknowledged—both because recognition of historical injustice and its legacies is vital for securing just relations in the future, and because our efforts to reduce vulnerability may backfire if its underlying causes are misidentified. As Cameron states in the case of the Inuit, 'if the very factors cited as undermining . . . capacities to adapt to climatic change are themselves a legacy of colonial intervention, then reframing . . . vulnerability as a matter of enhancing *local capacities*, rather than attending to the structural and systemic processes by which those capacities are continually undermined, must be challenged' (Cameron 2012, 110). Richard Howitt et al.—who describe colonization and marginalization as slower, underlying, 'unnatural disasters' that wreak havoc in indigenous communities—similarly argue that a just and sustainable response to climate disasters, which will support indigenous rights and resilience, 'requires acknowledgement that the outcome of natural disasters is often mediated by the unnatural disaster of colonial and post-colonial state policies and practices' (2012, 48). If we fail to acknowledge this, we appear to 'blame the victim' by suggesting that 'the problem rests in the inherent vulnerabilities and lack of capacity of indigenous people and their culture' (Howitt et al. 2012, 57).

The link between historical injustice and present-day liability to bear the *costs* of climate change is, however, harder to make out. For a start, the relative causal efficacy of any given factor that has contributed to climate risk will be difficult to determine. Then, as Shue points out, 'causal responsibility does not translate smoothly into moral responsibility' (Shue 1992, 391, fn. 11); some of those causally implicated in climate risk may not be morally culpable (e.g., individuals and collectives that have contributed to exposure through blameless decisions to form settlements in coastal areas). And finally, to add further complication, both causal and moral responsibility for climate risk may fail to ground any present liabilities to bear the costs of climate adaptation or compensation for loss and damage. Those whom we judge to be morally or causally responsible may turn out to no longer be alive, or there may be reasons to excuse them from bearing the costs (because they are impoverished, for example). The problem of dead emitters is sometimes addressed by appealing to a beneficiary pays principle, but on the current—more complex—picture it is not clear which beneficiaries we should assign liability *to*: how should we share costs between the beneficiaries of GHG emissions, deforestation, fossil fuel extraction and sale,[12] or colonial and other practices that have contributed to vulnerability and exposure?

Thus, even where we can conclude that certain historical injustices have causally contributed to climate vulnerability, it will be difficult to use this conclusion to determine precisely who should bear the costs of preventing or responding to any resulting climate harms. Furthermore, it is important to be clear about what this link between climate change and historical injustice *does not* show. I certainly do not intend to imply that the legacies of historical injustice should only be rectified if they are now contributing to the problem of climate change; nor do I intend to suggest that socio-economic circumstances that place communities at risk, or undermine our attempts to deal with climate change fairly, should only be addressed if they are the result of historical injustice.

Given the urgent nature of the problem, the best that we may be able to do currently is assign the costs of dealing with climate change on the basis of ability to pay. A more principled allocation of costs appears to depend on difficult determinations of causal responsibility and moral culpability, making the ability to pay approach a good pragmatic alternative. One may then wonder what exactly is gained from the attention to history that I have been advocating. Isolationists about climate justice (who treat climate change as an issue that can be dealt with independently of other matters of global justice)[13] are already likely to hold that past wrongs should simply be ignored in our theorising about climate justice for reasons of simplicity and feasibility. Axel Gosseries, for example, chooses to proceed on the assumption that 'the allocation of territories and natural resources among countries was a fair one (although we know that it is not)' (2005, 283). Eric Posner and David Weisbach—who concede that 'some of the world's most serious problems' include an unfair global distribution of wealth and 'the lingering harms of the legacy of colonialism'—similarly suggest that treating the climate negotiations as 'an opportunity' to solve these problems will only be counterproductive, and argue that we should therefore ignore such matters in the formulation of a climate treaty (Posner and Weisbach 2010, 5). By highlighting how difficult it is to determine present-day climate liabilities in a way that takes past wrongs into account, my discussion may simply seem to succeed in offering further support for such pragmatic disregard of history.

The problem with failing to attend to history, however, is that we will then also fail to recognise how climate change is actually part of an on-going, or enduring, injustice. Our present circumstances may be notable in that the threats resulting from global natural resource use have become, fairly quickly, unprecedented in scope and magnitude. However, many of the underlying vulnerabilities that the physical phenomenon of climate change interacts with in producing harm have been present for far longer, and derive from similar practices of exploiting natural resources at the expense of important human interests. Only when an injustice is on-going in this way are we confronted with the question of *why* it endures; and when this question arises, we are given reason to search for persisting structures that perpetuate injustices and prevent their rectification. Thus, as Iris Marion Young says, 'an account of the continuities of present with past injustices' can be important 'for understanding how the present conditions are structural, how those structures have evolved, and where intervention to change them may be most effective' (2011, 181–82).[14]

Following Young, we can understand a structural injustice to exist 'when social processes put large groups of persons under systematic threat of domination or deprivation of the means to develop and exercise their capacities, at the same time that these processes enable others to dominate or to have a wide range of opportunities for developing and exercising capacities available to them' (2011, 52). One such social process that appears to have persisted beyond the colonial period to our current circumstances of climate change results from the global system of natural resource governance; a system in which the benefits of natural resources tend to accrue disproportionately to the wealthy and powerful, putting large numbers of resource-dependent people at systematic threat of domination and deprivation.

One important aspect of the wrong of colonialism was the widespread and long-standing practice, by colonial authorities, of governing and exploiting natural resources in ways that dominated and deprived the local population. Though colonialism is now supposed to be a thing of the past—with international law affirming 'the right of peoples and nations to permanent sovereignty over their natural wealth and resources' (UN 1962, Art.1)—many collectives worldwide continue to find their jurisdiction over territorial resources undermined by forces beyond their control. A major example of this continuing social process of expropriation is the international resource privilege.[15] This global market rule—upheld by (generally wealthy) importing states—effectively hands the legal right to sell off the resources of a given territory to whoever can maintain coercive control over the local population and, as Wenar suggests, can plausibly be seen 'as a holdover from an earlier era of expansive sovereignty and colonial rule' (2008, 14).

The resource privilege can have a severe impact on the domestic arrangements of resource-rich countries. In a phenomenon known as the 'resource curse', the problematic incentive and power structures created by this privilege undermine democracy and economic growth and support authoritarian rule and civil conflict, thus preventing the people of a resource cursed country from exercising any meaningful form of collective self-determination. The international actors that uphold the system engendering the resource curse—whether by endorsing the resource privilege, trading with unjust regimes, or even installing and propping up authoritarian rulers that are willing to sell off local resources at acceptable prices—are thereby implicated in a severe injustice of global resource governance.[16]

The resource privilege is perhaps the most obvious example of the way in which the global system of natural resource governance remains—to a large extent—a system of might makes right, reminiscent of the colonial period. This is a system in which various structures (including laws, incentives and markets) serve to ensure that the rich and powerful can access and control the resources that they want, whilst poor and vulnerable collectives are often unable to make decisions about the resources on which they live and depend—or to resist displacement, extraction and expropriation.[17]

It is important to recognise this structural injustice because if we do not address these problematic structures of resource governance, then there is good reason to fear that even well intentioned efforts to find a fair solution to the problem of climate change will be undermined. As W. Neil Adger states, though 'climate change is a significant challenge to structures of governance at all temporal and spatial scales', this is particularly so 'in the area of managing natural resources' (2001, 921). One thing that the international climate negotiations will do, in effect, is create new patterns of control over the world's resources: in particular the global GHG sink, but also fossil fuels, forests, land, water, and various other natural resources that can be used in offsetting schemes or the production of renewable energy. If persisting structures of injustice in global resource governance are not addressed, the climate regime threatens to become another way by which problematic inequalities can be perpetuated and even enhanced, with parties that occupy a position of dominance further expanding their control over the Earth's resources at the expense of the globally disadvantaged.

Thus, to take a more integrated approach to climate justice—considering the role that historical wrongs have played in bringing it about—is not to use the climate negotiations as 'an *opportunity*' to solve these problems (Posner and Weisbach 2010, 5; emphasis added). Rather, it is to adopt a perspective that enables us to identify persisting structural injustices—such as those regarding the way that the Earth's resources are used, shared and controlled—and to recognise how certain forms of climate governance may serve to perpetuate such problems.

## WHAT DOES THIS MEAN FOR CLIMATE GOVERNANCE?

The problem of structural injustice identified in the previous section should create concerns about what Matthew Paterson has described as 'the marketization of climate governance' (2011, 617). Paterson here refers to Peter Newell's discussion of the marketization of environmental governance more generally (see Newell 2008), what Newell refers to as 'an ensemble of strategies of market governance including practices of privatisation and commodification of natural resources which derive from a common belief in the ability of markets to provide the public good of environmental protection in the most efficient way' (2005, 189). In the climate change case, Paterson suggests, such marketization can be observed in the 'major trend in the international climate regime . . . towards the organization of climate governance through the creation of markets in rights to emit GHGs' (2011, 615) and can be explained by the ability of such markets to 'create concentrated, immediate benefits for powerful actors' (Paterson 2011, 620).

The main worry regarding such marketization is that market solutions are likely to favour parties that are already unjustly advantaged if not designed carefully. As Simon Caney and Cameron Hepburn note: 'In general, market systems have a tendency at best to perpetuate existing distributions of wealth, and at worst to exacerbate wealth differences between rich and poor' (2011, 223). Creating new, economic value in the natural resources on which communities depend will not necessarily benefit those communities in a world where access to natural resources tends to be determined by wealth and power, and where many poor people still struggle to realise their resource rights due, for example, to insecure land tenure arrangements and exclusion from environmental decision-making. Instead, it could just create new vulnerabilities: to exploitation by outside agents—or agents of the state—who may then seek to seize those resources in order to obtain the economic benefits for themselves.

'Global carbon pricing', for example, threatens to exacerbate existing inequalities of control over the global GHG sink, creating a system of governance in which 'rich and poor states could not possibly participate on fair and equal terms since the former could draw on their superior financial resources to emit far more greenhouse gas than the latter' (Page 2013, 243). Similar concerns are raised by market measures like the Clean Development Mechanism and the UN's REDD and REDD+ mitigation schemes, which 'act on the principle of industrialized countries (or those who can pay) offsetting their effluents by investing in the developing world' (Marino and Ribot 2012, 324). These measures are designed to place climate mitigation projects in some of the poorest regions globally, and thus threaten to expose already vulnerable communities to any potentially harmful side effects of such developments. REDD and REDD+, for example, offer financial rewards for developing countries that reduce deforestation and forest degradation (see UN 2009). Some worry that this new form of forest governance may restrict local access to resources, impinge on local livelihoods and dispossess resident communities (Cotula and Mayers 2009, 3; Larson 2011, 547). Such schemes also threaten to recentralise governance by placing forests under state control, potentially undermining the participation of local communities in decision-making about their environment (Phelps et al. 2010).

Historical injustices of natural resource governance thus have present moral significance because they are part of an enduring injustice in global resource governance that persists to the present day, and which now threatens to prevent us from dealing with climate change fairly. Instead of allowing market processes to determine access to and control over the world's resources, we should be ensuring that climate governance recognises the agency of vulnerable communities, directly strengthens their decision-making power, and protects their ability to control the land and natural resources on which they depend.

## CONCLUSION

Though the history of natural resource use is often claimed to be relevant to our attempts to deal with climate change fairly, few theorists concerned with historical use of the climate sink have, thus far, expanded their attention to consider the way in which broader historical methods of natural resource governance—the unjust governance that took place during periods of colonialism, for example—could also possess contemporary moral significance. The problem of historical injustice in the use of the Earth's resources is much broader than the problem of climate change; but it is relevant insofar as it has made climate change a particularly challenging problem of global justice, in which communities are differentially vulnerable to climate events and persisting injustices of control over and access to natural resources that threaten to undermine fair climate governance. Acknowledging and addressing this problem is vital and urgent if we are to ensure that the climate negotiations—and the new systems of resource governance that they are now in the process of creating—do not perpetuate such injustice.

## REFERENCES

Acemoglu, Daron. 2003. 'Root Causes'. *Finance & Development*: 27–30.

Acemoglu, Daron, and James A. Robinson. 2010. 'Why Is Africa Poor'? *Economic History of Developing Regions* 25 (1): 21–50.

———. 2012. *Why Nations Fail.* New York: Crown Publishers.

Acemoglu, Daron, Simon Johnson, and James A. Robinson. 2001. 'The Colonial Origins of Comparative Development: An Empirical Investigation'. *The American Economic Review* 91 (5): 1369–1401.

———. 2002. 'Reversal of Fortune: Geography and Institutions in the Making of the Modern World Income Distribution'. *The Quarterly Journal of Economics* 117 (4): 1231–94.

———. 2006. 'Understanding Prosperity and Poverty: Geography, Institutions, and the Reversal of Fortune'. In *Understanding Poverty*, edited by Abhijit Vinayak Banerjee, Roland Bénabou and Dilip Mookherjee, 19–35. Oxford: Oxford University Press.

Adger, W. Neil. 2001. 'Scales of Governance and Environmental Justice for Adaptation to Climate Change'. *Journal of International Development* 13: 921–31.

Athanasiou, Tom, and Paul Baer. 2002. *Dead Heat: Global Justice and Global Warming.* New York: Seven Stories Press.

Barnett, Jon, and John Campbell. 2010. *Climate Change and Small Island States: Power, Knowledge, and the South Pacific*. London: Earthscan.

Beckerman, Wilfred, and Joanna Pasek. 1995. 'The Equitable International Allocation of Tradable Carbon Emission Permits'. *Global Environmental Change* 5 (5): 405–13.

Bell, Derek. 2010. 'Justice and the Politics of Climate Change'. In *Routledge Handbook of Climate Change and Society*, edited by Constance Lever-Tracy, 423–41. London: Routledge.

Butt, Daniel. 2009a. *Rectifying International Injustice: Principles of Compensation and Restitution Between Nations*. Oxford: Oxford University Press.

———. 2009b. ''Victors' Justice'? Historic Injustice and the Legitimacy of International Law'. In *Legitimacy, Justice and Public International Law*, edited by Lukas H. Meyer, 163–85. Cambridge: Cambridge University Press.

———. 2012. 'Repairing Historical Wrongs and the End of Empire'. *Social & Legal Studies* 21 (2): 227–42.

———. 2013. 'Colonialism and Postcolonialism'. In *The International Encyclopedia of Ethics,* edited by Hugh LaFollette, 892–98. Malden, MA: Wiley-Blackwell.

Cameron, Emilie S. 2012. 'Securing Indigenous Politics: A Critique of the Vulnerability and Adaptation Approach to the Human Dimensions of Climate Change in the Canadian Arctic'. *Global Environmental Change* 22: 103–114.

Caney, Simon. 2006. 'Environmental Degradation, Reparations, and the Moral Significance of History'. *Journal of Social Philosophy* 37 (3): 464–82.

———. 2012. 'Just Emissions'. *Philosophy & Public Affairs* 40 (4): 255–300.

Caney, Simon, and Cameron Hepburn. 2011. 'Carbon Trading: Unethical, Unjust and Ineffective?' *Royal Institute of Philosophy Supplement* 69: 201–34.

Cotula, Lorenzo, and James Mayers. 2009. 'Tenure in REDD—Start-Point or Afterthought?' *Natural Resource Issues* 15. London: International Institute for Environment and Development.

Duus-Otterström, Göran. 2014. 'The Problem of Past Emissions and Intergenerational Debts'. *Critical Review of International Social & Political Philosophy* 17 (4): 448–69.

Figueroa, Robert Melchior. 2011. 'Indigenous Peoples and Cultural Losses'. In *The Oxford Handbook of Climate Change and Society*, edited by John S. Dryzek, Richard B. Norgaard and David Schlosberg, 232–47. Oxford: Oxford University Press.

Gardiner, Stephen M. 2011. *A Perfect Moral Storm: The Ethical Tragedy of Climate Change*. Oxford: Oxford University Press.

Gosseries, Axel. 2005. 'Cosmopolitan Luck Egalitarianism and the Greenhouse Effect'. *Canadian Journal of Philosophy* 35, Supplementary Volume 31: 279–309.

Grubb, Michael, James Sebenius, Antonio Magalhaes, and Susan Subak. 1992. 'Sharing the Burden'. In *Confronting Climate Change: Risks, Implications and Responses*, edited by Irving M. Mintzer, 305–22. Cambridge: Cambridge University Press.

Halme, Pia. 2007. 'Carbon Debt and the (In)Significance of History'. *Trames* 4: 346–65.

Howitt, Richard, Olga Havnen, and Siri Veland. 2012. 'Natural and Unnatural Disasters: Responding with Respect for Indigenous Rights and Knowledges'. *Geographical Research* 50 (1): 47–59.

IPCC. 2007a. *Climate Change 2007: Impacts, Adaptation and Vulnerability. Contribution of Working Group II to the Fourth Assessment Report of the Intergovernmental Panel on Climate Change*. Edited by M. L. Parry, O. F. Canziani, J. P. Palutikof, P. J. van der Linden, and C. E. Hanson. Cambridge, UK, and New York, NY: Cambridge University Press.

———. 2007b. *Climate Change 2007: Synthesis Report. Contribution of Working Groups I, II and III to the Fourth Assessment Report of the Intergovernmental Panel on Climate Change.* Edited by R. K Pachauri and A. Reisinger. Geneva, Switzerland: IPCC.

———. 2012. *Managing the Risks of Extreme Events and Disasters to Advance Climate Change Adaptation. A Special Report of Working Groups I and II of the Intergovernmental Panel on Climate Change.* Edited by C. B. Field, V. Barros, T. F. Stocker, D. Qin, D. J. Dokken, K. L. Ebi, M. D. Mastrandrea, K. J. Mach, G.-K. Plattner, S. K. Allen, M. Tignor, and P.M. Midgley. Cambridge, UK, and New York, NY: Cambridge University Press.

———. 2014a. *Climate Change 2014: Impacts, Adaptation, and Vulnerability. Part A: Global and Sectoral Aspects. Contribution of Working Group II to the Fifth Assessment Report of the Intergovernmental Panel on Climate Change.* Edited by C. B. Field, V.R. Barros, D.J. Dokken, K.J. Mach, M.D. Mastrandrea, T.E. Bilir, M. Chatterjee, K.L. Ebi, Y.O. Estrada, R.C. Genova, B. Girma, E.S. Kissel, A.N. Levy, S. MacCracken, P.R. Mastrandrea, and L.L. White. Cambridge, UK, and New York, NY: Cambridge University Press.

———. 2014b. *Climate Change 2014: Impacts, Adaptation, and Vulnerability. Part B: Regional Aspects. Contribution of Working Group II to the Fifth Assessment Report of the Intergovernmental Panel on Climate Change.* Edited by V. R. Barros, C.B. Field, D.J. Dokken, M.D. Mastrandrea, K.J. Mach, T.E. Bilir, M. Chatterjee, K.L. Ebi, Y.O. Estrada, R.C. Genova, B. Girma, E.S. Kissel, A.N. Levy, S. MacCracken, P.R. Mastrandrea, and L.L. White. Cambridge, UK, and New York, NY: Cambridge University Press.

Kartha, Sivan. 2011. 'Discourses of the Global South'. In *The Oxford Handbook of Climate Change and Society*, edited by John S. Dryzek, Richard B. Norgaard, and David Schlosberg, 504–18. Oxford: Oxford University Press.

Larson, Anne M. 2011. 'Forest Tenure Reform in the Age of Climate Change: Lessons for REDD+'. *Global Environmental Change* 21: 540–49.

Lu, Catherine. 2011. 'Colonialism as Structural Injustice: Historical Responsibility and Contemporary Redress'. *Journal of Political Philosophy* 19 (3): 261–81.

Marino, Elizabeth, and Jesse Ribot. 2012. 'Special Issue Introduction: Adding Insult to Injury: Climate Change and the Inequities of Climate Intervention'. *Global Environmental Change* 22 (2): 323–28.

Martinez-Alier, Joan, and Stephen Naron. 2004. 'Ecological Distribution Conflicts and Indicators of Sustainability'. *International Journal of Political Economy*, 34 (1): 13–30.

Meyer, Lukas H., and Dominic Roser. 2006. 'Distributive Justice and Climate Change. The Allocation of Emission Rights'. *Analyse & Kritik* 28: 223–49.

Miller, David. 2008. 'Global Justice and Climate Change: How Should Responsibilities Be Distributed?' The Tanner Lectures on Human Values, delivered at Tsinghua University, Beijing, 24–25 March: 119–56. http://www.tannerlectures.utah.edu/lectures/documents/Miller_08.pdf.

Nagel, Thomas. 2005. 'The Problem of Global Justice'. *Philosophy & Public Affairs* 33 (2): 113–47.

Neumayer, Eric. 2000. 'In Defence of Historical Accountability for Greenhouse Gas Emissions'. *Ecological Economics* 33 (2): 185–92.

Newell, Peter. 2005. 'Towards a Political Economy of Global Environmental Governance'. In *Handbook of Global Environmental Politics*, edited by P. Dauvergne, 187–201. Cheltenham: Edward Elgar.

———. 2008. 'The Marketization of Global Environmental Governance'. In *The Crisis of Global Environmental Governance*, edited by J. Park, K. Conca, and M. Finger, 77–95. London: Routledge.

Page, Edward. 2012. 'Give It Up For Climate Change: A Defence of the Beneficiary Pays Principle'. *International Theory* 4 (2): 300–30.

———. 2013. 'Climate Change Justice'. In *The Handbook of Global Climate and Environment Policy*, edited by Robert Falkner, 231–47. Malden, MA: Wiley-Blackwell.

Paterson, Matthew. 2011. 'Selling Carbon: From International Climate Regime to Global Carbon Market'. In *The Oxford Handbook of Climate Change and Society*, edited by John S. Dryzek, Richard B. Norgaard, and David Schlosberg, 611–24. Oxford: Oxford University Press.

Phelps, Jacob, Edward L. Webb, and Arun Agarwal. 2010. 'Does REDD+ Threaten to Recentralize Forest Governance?' *Science* 328: 312–13.
Pickering, Jonathan, and Christian Barry. 2012. 'On the Concept of Climate Debt: Its Moral and Political Value'. *Critical Review of International Social & Political Philosophy* 15 (5): 667–85.
Pogge, Thomas W. 2002. *World Poverty and Human Rights: Cosmopolitan Responsibilities and Reforms*. Cambridge: Polity Press.
Posner, Eric A., and David Weisbach. 2010. *Climate Change Justice*. Princeton; Oxford: Princeton University Press.
Risse, Mathias. 2005. 'Do We Owe the Global Poor Assistance or Rectification?' *Ethics & International Affairs* 19 (1): 9–18.
———. 2012. *On Global Justice*. Princeton: Princeton University Press.
Roberts, J. Timmons, and Bradley C. Parks. 2006. 'Globalization, Vulnerability to Climate Change, and Perceived Injustice'. *Society & Natural Resources: An International Journal* 19 (4): 337–55.
———. 2007. *A Climate of Injustice: Global Inequality, North-South Politics, and Climate Policy*. Cambridge, MA: MIT Press.
Shue, Henry. 1992. 'The Unavoidability of Justice'. In *The International Politics of the Environment*, edited by Andrew Hurrell and Benedict Kingsbury, 373–97. Oxford: Clarendon Press.
Sinden, Amy. 2010. 'Allocating the Costs of the Climate Crisis: Efficiency Versus Justice'. *Washington Law Review* 85: 293–353.
Smith, Kirk R. 1991. 'Allocating Responsibility for Global Warming: The Natural Debt Index'. *Ambio* 20 (2): 95–96.
Spinner-Halev, Jeff. 2012. 'Historical Injustice'. In *The Oxford Handbook of Political Philosophy*, edited by David Estlund, 319–22. Oxford: Oxford University Press.
UN. 1962. 'Permanent Sovereignty Over Natural Resources'. General Assembly Resolution 1803 (XVII), December 14. Accessed April 27, 2014. http://www.ohchr.org/EN/ProfessionalInterest/Pages/NaturalResources.aspx.
———. 2009. 'UN-REDD Programme'. The United Nations Collaborative Programme on Reducing Emissions from Deforestation and Forest Degradation in Developing Countries. Accessed January 25, 2014. http://www.un-redd.org/.
Wenar, Leif. 2008. 'Property Rights and the Resource Curse'. *Philosophy & Public Affairs* 36 (1): 2–32.
Young, Crawford. 1994. *The African Colonial State in Comparative Perspective*. New Haven, CT: Yale University Press.
Young, Iris Marion. 2011. *Responsibility for Justice*. New York: Oxford University Press.
Ypi, Lea, Robert E. Goodin, and Christian Barry. 2009. 'Associative Duties, Global Justice, and the Colonies. *Philosophy & Public Affairs* 37 (2): 103–35.

## NOTES

1. Early versions of this argument were presented at the ECPR Joint Sessions of Workshops 2014 and the ALSP Annual Conference 2014, and I am grateful to the organisers and attendees of those events for their helpful feedback. Thanks are also due to Bevan Richardson, for many discussions on this topic that helped me to straighten my thoughts; Chris Bertram, Joanna Burch-Brown, Simon Caney, Fabian Schuppert and participants of the Stanford postdoc seminar for their comments on previous written versions; and Catriona McKinnon and Aaron Maltais for helping me to significantly improve the chapter.

2. The 'natural debt' terminology is used in Grubb et al. 1992, 312; Meyer and Roser 2006, 238; Neumayer 2000, 186; and Smith 1991. Alternative terms that are commonly invoked during discussions of climate change and historical responsibility include 'ecological debt', 'climate debt', 'carbon debt', and—more rarely—'atmospheric debt'. See the discussions in: Athanasiou and Baer 2002, 121; Duus-Otterström 2014, 450; Halme 2007; Kartha 2011, 508-9;

Martinez-Alier and Naron 2004, 19; Pickering and Barry 2012; Risse 2012, 394, fn.16; and Sinden 2010.

3. For critiques of the historical emissions debt view see Beckerman and Pasek 1995, 410; Caney 2006; Miller 2008, 133–37; and my paper titled 'Historical Use of the Climate Sink'.

4. See Spinner-Halev's discussion of what he terms 'enduring injustice' (2012, 329).

5. One obvious example of such burdens being the duty to restrict GHG emissions.

6. The BPP—defended by Bell (2010, 437–38) and Page (2012), among others—assigns the costs of dealing with climate change to those who have benefitted from excessive emissions of GHGs.

7. Page, for example, argues that the benefits and burdens of historical use of GHG sinks should be distributed in a compensatory manner because they 'share common origins' (Page 2012, 313). I criticise this position in my paper titled 'Historical Use of the Climate Sink'.

8. Roughly speaking, the IPCC understands a disaster to be an adverse impact which 'produce[s] widespread damage and cause[s] severe alterations in the normal functioning of communities or societies' (IPCC 2012, 4).

9. Gardiner claims that our present circumstances of global injustice are also a result of 'currently pronounced global poverty and inequality, and the role of rich nations in structuring existing transnational institutions' (2011, 119). Later on, I will suggest that colonialism in fact played a significant role in creating such poverty and inequality, and in ensuring that transnational institutions were set up according to the interests of the richer nations.

10. Such as 'mining and lumbering resources as well as ranching and plantation agriculture' (Roberts and Parks 2007, 112).

11. In another paper, Roberts and Parks draw on three case studies to argue that 'the "root causes" of climate disasters lie in colonial histories and current relations with the global economy that keep these nations vulnerable' (2006, 351).

12. In many instances, the case for deeming the beneficiaries of fossil fuel extraction and sale both culpable and liable for the costs of climate change has been enhanced by their efforts to undermine climate mitigation efforts.

13. For more on the distinction between isolationism and integrationism about climate justice, see Caney 2012, 258–59.

14. As Catherine Lu points out in her own discussion of the lasting effects of colonialism: 'there is a distinction between acts of injustice being past and structural injustice being a thing of the past. Even if unjust acts or policies end . . . unjust structural processes and conditions may persist' (2011, 278).

15. This privilege has received a fair amount of attention in the philosophical literature. See Pogge 2002, 112–14, 162–66; Wenar 2008.

16. Furthermore, this injustice appears to have additional relevance to the problem of climate change. Fair global governance of fossil fuels will be essential for dealing with climate change justly; but oil is one of the major natural assets associated with the resource curse. By encouraging authoritarian rule in oil-rich regions, the international resource privilege has thereby created a significant stumbling block to effective global action on climate change—a stumbling block for which we may wish to hold certain international actors responsible.

17. Cameron similarly identifies 'lack of control over resource development' as a persisting difficulty faced by the Inuit that is generally understood 'in relation to colonization' (2012, 106).

*Chapter Two*

# International Domination and a Global Emissions Regime

Patrick Taylor Smith[1]

Much philosophical ink has been spilled concerning the question of what *principles* we should use to distribute rights or entitlements to emit greenhouse gases (Posner & Weisbach 2013; Caney 2012; Blomfield 2013; Jameison 2005; Neumayer 2000; Vanderheiden 2008; Gosseries 2011). What has received significantly less attention is the question of how the regime that will actually distribute these rights ought to be constructed. Part of the reason for this failure is that many theorists treat the construction of governance regimes purely instrumentally: a regime is good insofar as it reliably brings about the satisfaction of the relevant principles of justice. Yet, it is plausible to think that the justification of political institutions depends upon not only what they do but also on *how* they do it. One might think that a benevolent dictator relates unjustly to her subjects even if she perfectly follows the correct principles of justice in her administration. That is, one is unfree when one is subject to a dictator or slaveowner even if he furthers my interests. With that in mind, I plan to justify the need for a global emissions regime, describe the features of that regime, and lay out some problems that regime will face and need to eventually resolve from the standpoint of freedom as non-domination.

Three significant conclusions follow from this focus on freedom as non-domination when it is combined with particular facts about climate change. First, a governance regime that relies on unilateral state action—even when effectively coordinated—is inadequate: the regime will need to be a formal organization with significant independence and authority. Second, while we might be able to develop emissions *principles* while only considering issues concerning climate change (Posner & Weisbach 2013), it is impossible for non-dominating environmental governance to be so limited. That is, even if we grant that emissions principles could be characterized independently of other considerations, this chapter shows that any plausible attempt at the rightful enforcement of those principles will fail in the face of egregious global poverty and lack of development. Third, once we create an independent, authoritative regime with the needed power, then we will need to ensure that *its* power is non-dominating and this will be unusually difficult; so difficult, in fact, that perhaps the most we can achieve is minimization of the risks of domination, rather than non-domination per se.

Climate change intersects in an especially problematic way with the geopolitical realities that currently govern the international system. This is because, generally speaking, differences in *vulnerability* track differences in *power*. Those states that emit the most are, non-accidentally, the most powerful and richest in the system. This connection is the result of a complex admixture of factors. First, geopolitical power is, at its base, partially dependent on a society's economic growth and productivity, which in turn has historically been produced by industrial processes and structures that have been reliant on the burning of fossil fuels and consequent greenhouse gas emissions. Second, powerful states have structured the international system in ways that have ensured their access to fossil fuels and other resources on terms favorable to themselves while simultaneously burdening and exploiting weaker nations.[2] Also, many of these states are geographically situated so as to be less vulnerable to climate change impacts. These natural features—greater security in terms of water and food, for one example—are greatly amplified by their greater wealth; they spend more on adapting to climate change impacts, reducing the negative effects on their own citizenry. Conversely, many of those countries that emit the least are among the poorest and least powerful in the system while also possessing naturally less resilient ecologies, more insecure food systems or vulnerable low-lying coastal areas. The greater geographical vulnerability of those countries—as well as the concentration of the worst impacts of climate change—is, again, amplified by the fact that their poverty prevents them from easily responding to those greater impacts (Vanderheiden 2008; Posner & Weisbach 2013; IPCC 2014). And these tendencies can be added to a game theoretical element: those nations with the greatest incentive to free ride are those nations that have the greatest power and resources to do so (Gardiner 2002). So geography, wealth, power and the incentive structure of the current, business-as-usual emissions regime all combine to create an alignment of interests amongst high emitters to avoid responding effectively to climate change impacts, providing ample opportunities for domination, exploitation and oppression.

## POWER, DOMINATION AND FREEDOM

Global environmental governance will, by necessity, involve a restructuring of the political entitlements held by individuals. Under the status quo, particular states have broad discretionary authority to decide their own emissions policy and, as a consequence, particular individuals have coercively backed entitlements to emit carbon irrespective of broader social cost. For example, when I commute to and from work, the state coercively protects my right to pay the current market price for gasoline without compensating for the negative externalities of that purchase. If an international body, an NGO, or another state attempted to subject me to coercive sanctions, my state would presumably deploy force in order to protect my entitlement (or would at least consider itself entitled to do so). Conversely, a new environmental governance regime would remove that final, discretionary authority from current states and their citizens while distributing a new set of legal and political powers and entitlements to a different group of agents.[3] As a consequence of this redistribution, individuals will be differentially subject to political authority in ways that will restrict their ability to pursue their projects, satisfy their desires and enact their conception of the good.[4] Of course, the *current* system also does this; the distribution of current entitlements similarly constrains some and empowers others. In other words, the political, social and economic institutions and norms that make up 'business as usual' or 'effective climate governance' will both generate and reflect power relations, and those power relations demand justification. This chapter is primarily concerned with the worry that, in the process of reacting to the real governance failures of the current state system in responding to climate change, we will generate a new set of power relations that are themselves problematic. That is, in our need to respond to the way our current systems of political authority unjustifiably fail to address the externalities associated with GHG emissions, we might end up recapitulating the same sorts of problematic political relations. And this worry is deepened by the fact that truly effective climate governance will need to be quite powerful.

In order for us to even understand this worry—much less respond to it—we need an account of what it is for power relations to be rightful. I will suggest that the best way to understand rightful relations is by reference to the concept of domination. Claiming that one is dominated is, on this view, (Lovett 2012) a normative complaint that can be aimed at political institutions in their possession and exercise of power. In other words, the core idea is that being dominated makes one unfree and that the unfreedom is a deep injustice that demands resolution. (See Pettit 1997, 2013; Ripstein 2014, 2009; Young 2006.)

Yet, this is only to say that it is vitally important to have political relations that ensure that everyone is free of domination. I have not said what domination consists of and how it can be avoided. Domination, in this view, is a result of how two agents relate to one another: X is dominated by Y when X is subjected to the arbitrary superior power of Y. So the two key elements of this domination are 1) superior power and 2) arbitrariness. Let's take each in turn. Power is the capacity to structure the choices and life chances of another, and its sources are both contextual and diverse. Usually, one's social position—often defined by different sorts of privileges, rights and entitlements—is the source of one's superior power. For example, modern police officers, especially while operating in their official capacities as agents of the state, have superior power over almost all citizens in virtue of training, technology and social organization. Turning to arbitrariness, the key feature of the republican-domination view is that merely having a good reason for acting does not mean that one is acting non-arbitrarily. What's more, adopting the right principle—even if one is reliably disposed to follow it—also does not necessarily make one's power non-arbitrary. Rather, non-arbitrariness is a function of the actual relations of accountability, contestation and control that constrain, regulate and structure the deployment and exercise of that power for particular ends. So, the police in a well-ordered liberal democracy do not dominate when there are institutional constraints, checks and balances and other mechanisms for ensuring that they exercise their power in a way that makes them accountable to the claims of those subject to it. These constraints may take different forms: civilian review boards with the authority to sanction police officers for abuses, civilian control that is subject to elections, an independent judiciary, or training the professionalizers so that they constrain each other. In the republican view, these tools do not merely make it less likely that the police will abuse their power. Rather, they help constitute a different kind of political relation, a relation that allows the police to deploy power for the common good in a way consistent with everyone's freedom. Similarly, if the police happen to serve the common good without these tools, this does not change the fact that their power is still dominating. A benevolent despot remains a despot. Of course, it is, in some sense, better to be under the heel of a kind tyrant than a cruel one, but it is still true that no one should be under any person's heel.

## INTERNATIONAL DOMINATION AND CLIMATE CHANGE: FROM BUSINESS AS USUAL TO GLOBAL GOVERNANCE

International politics seems to be especially likely to produce domination. Generally speaking, variations in intrinsic human capabilities are relatively small. As a consequence, a significant proportion of an individual's power is a consequence of the political, social and economic order within which they find themselves. So, it is a comparatively simple matter of altering power relations by changing the social structures that generate the differences in power in the first place. In the international realm, however, states possess very large differences in intrinsic capability and those differences are *not* almost entirely a consequence of the particular structure of international politics.[5] Moreover, international institutions are comparatively inchoate or weak; they are less capable of responding to or compensating for these large differences in intrinsic capabilities. Finally, many of the standard strategies that have been developed in dealing with the possibility of domestic domination appear to presuppose a common constitutional order, shared institutions and, more generally, the notion of a self-determining, collective *demos*. In the international order, all of these conditions are absent, and many are skeptical that democratic or rule of law institutions can be readily 'scaled up' to deal with the global problems. These three features—large differences in intrinsic capability, weak governing institutions and the lack of a common demos—deepen the possibilities for domination and complicate any efforts to create rightful relations in the face of large differences of power.

All of these issues apply to the status quo of climate change governance. States vary greatly in how much they can and do emit, in how much they can adapt to climate change impacts, in how readily they can mitigate their own emissions and in how much they will suffer from climate change impacts. These differences—combined with more general differences in economic and political power—generate large differences in the bargaining position of states when it comes to negotiating a new emissions regime. Second, international institutions are currently very weak in terms of regulating emissions behaviour. Third, there is still little sense of a common, global demos when it comes to environmental issues, though perhaps one can see the beginnings of an environmental and globally minded civil society. Yet, there is one additional complicating factor. Unlike other issues where we might imagine that states with common interests might band together against powerful yet ill-behaved actors, there is an unfortunate connection between emissions behaviour and intrinsic power. The states most responsible for climate change are the powerful actors in the system.[6] Those states that will suffer the most tend to be those states that are economically and politically weak. So, it is unlikely that a balancing coalition in order to generate serious costs for the primary emitters will form and be effective.

Current emissions behaviour by rich and powerful countries represents a serious case of international domination (Smith 2013) that does and will produce deeply problematic effects on the life chances of the poor and the weak. By emitting carbon at the business as usual rate, emitters structure the life chances and choices available to others. The people of Bangladesh are going to lead lives of greater hardship, less development and will be forced to expend scarce resources to respond to climate change impacts due to the emissions decisions of powerful countries. There are, as we shall see, no adequate mechanisms of accountability or contestation that allow the people of Bangladesh to call the emitters of China, the United States, or Australia into account.[7] It is precisely these facts that make it possible for these states to emit without consequences. Of course, these high emitters may realize that the costs of their behaviour will be sufficiently high *for them* and they decide to curb their behaviour, but this is does not change the relations of political accountability between high and low emitters. The large-scale unilateral action that has been proposed by both the United States and China might make the domination more palatable—assuming that such action reduces climate change impacts—but it does not fundamentally alter the political relations regulating the behaviour.[8] Just as a master can sincerely care for a slave and decide to give her free time, gifts, or education while still dominating her, the people and leaders of China and the United States can sincerely care about the interests of the people who will suffer egregiously as a result of climate change, act to reduce those impacts, and yet still dominate. This is because both the United States and China remain in the position of deciding for themselves the extent to which they reduce these climate impacts rather than a global political order that allows those decisions to be accountable to those who are subject to them. So, unilateral state action is not—even in principle—an attempt to resolve the dominating relationship evidenced by the essentially uncontested nature of business-as-usual state behaviour.[9] Other kinds of political structure are needed.

In what follows, I describe an environmental governance regime that will at least begin to resolve the problem of international domination in the context of climate change. I will begin with relatively minimal international institutions and show that preventing international domination will require granting additional power and authority to the regime. In the next section, I will then argue that the regime will have difficulties avoiding domination itself.

Let us begin with an international regime that serves primarily a *legislative* function, much like the current UNFCC. That is, the purpose of this institution is to create a deliberative body that will describe and promulgate rules that distribute a new set of entitlements to emit carbon and provide for the distribution of funds for adaptation to climate change impacts in order to facilitate swift movement towards sustainable global emissions behaviour. The point of this system would be to provide a set of coordinative principles.[10] One problem with many regimes of this sort is that they operate by state consensus. As a consequence, differences in power and bargaining position between states are recapitulated in the negotiating process. What's more, a consensus based decision procedure gives agents that are willing to walk away disproportionate power over the proceedings and, in contexts like climate change where the cooperation of the high emitters (who also represent the most powerful actors in the system), this often amounts to what is essentially a veto power. This, when combined with the facts about emissions that give a disproportionate incentive to walk away to those states that are going to be necessary for the effective functioning of the regime, grants too much power to high emitters. So, what are needed are compensatory mechanisms that equalize the bargaining position of the respective powers and a decision procedure that does not grant outsized influence to those states in stronger positions to walk away from an agreement.

Let's consider one such regime as an example. First, participation in the regime would be mandatory and binding even on actors that dissented; presumably, states would need to engage in a pre-commitment strategy by which attempts to exit the regime would be expected to fail. Second, voting power would be weighted according to anticipated total climate change impacts. So, states that had many people would—ceteris paribus—have more voting power, but those states who would suffer more severe impacts would receive weighted bonuses to their voting percentages. Nations like Bangladesh—which would suffer severe impacts and have considerable population—would find themselves with considerable more voting power than current systems—like the UN—where each state receives one vote. In any case, none of the subsequent arguments rely on the specific details of this regime. Rather, the key normative feature of these compensatory mechanisms is that, in the formulation of the legislative principles in the regime, they equalize the bargaining position of the relevant parties in order to ensure that the principles are formulated in a non-dominating fashion.

This purely legislative regime fails to resolve international domination for two reasons. First, it is a platitude that no legislative regime can cover all relevant cases (Hart 1961). There will be instances of state action or incorporation of the regime into domestic legal orders that will go beyond the principles, be subject to ambiguity, or be covered only vaguely. For example, imagine that the legislature requires that the monitoring of emissions—surely an issue of vital importance for determining whether the regime is or will succeed is preventing dangerous impacts—must be performed according to 'scientifically valid methods' to a 'reasonable degree of accuracy.' We can imagine the diversity of interpretations that each individual state—even in good faith—could adopt in light of those claims. And let us suppose that the regime purports to simply *list* the correct methods (and let's suppose, very implausibly, that all rules about the list, its application and the reporting of the results can be also specified) and let us further suppose that such an inflexible regime—in the face of the immense complexity of the climate change—would do a minimally adequate job in accomplishing the public policy goals of the regime. It would *still* be up to each state's discretion to determine whether they were using the methods on the list and, furthermore, each state could determine that there was a *penumbral* commitment that allowed them to use a different method. Under the purely legislative regime, these difficult or penumbral cases will be interpreted by the member agents, especially states. As a consequence, these judgements will result in domination. That is, the resolution of these cases—which will grant enormous flexibility to the states subject to the regime—will be unilaterally made by the states themselves.

So there are no mechanisms of accountability—in the purely legislative regime—that ensures that these judgements will be made in such a way as to be appropriately accountable to the interests of other states and takes their citizens into account. One might resist this conclusion in two ways. First, the legislative body could simply make more laws; legislative activity is not a one-off. Second, states would want to signal their cooperation in order to gain reciprocal cooperation from others, and this would serve as a constraint on state interpretation. I think there are several responses. First, we are now imagining a purely legislative regime, so I am assuming that the legislature will not be issuing specific injunctions to specific states in virtue of the body's determination that the rules have not been followed. So the regime would be limited to issuing general rules that purported to 'fix' the misinterpretations of member states, and this would likely be expensive and difficult with a judicial component. But even if it was not, without the ability to issue binding orders to specific states, states will always be in a position to issue their own interpretations of the rules. Second, the signaling objection recapitulates the point of the paragraph. Indeed, states will be limited by their need to achieve reciprocal cooperation to limit their interpretations to only as far as their bargaining position allows, but the whole point of creating the regime was to make it so that the relevant public policy outcomes did not depend on the bargaining power of the actors! So the United States will interpret the principles of the legislative regime as it applies to its own behaviour and laws, but how the United States interprets those cases will affect how it deploys its entitlements. This, in turn, will affect the life chances of the people of Bangladesh. The United States, in virtue of its superior position, will have a fair amount of discretion to make rulings in ways that benefits its citizenry at the expense of others. Bangladesh, subject to the power and influence of high emitting states, will have much less discretion. This asymmetry is a consequence of the superior power of high emitters and there is no mechanism for low emitters to contest or call into account these judgements. Of course, the high emitters might be conscientious and fair-minded. As a consequence, they may meticulously rule in ways that seem perfectly reasonable. Yet, it would remain up to the powerful to so rule, and there are no formal or material means for regulating or structuring that decision.[11]

This brings us to a possible reform of our proposed emissions governance regime. To our legislative regime, we should add a *judicial* component or function. The legislature could perform this function by, for example, designating specific subcommittees to hold hearings and issue injunctions. However, we would probably want courts to play somewhat of a counter-majoritarian and independent function, and I will assume that the judicial *function* should be performed by an independent yet related body. So, like the International Criminal Court (ICC) and the World Trade Organization (WTO), we should include an independent judicial body whose mandate was to interpret the principles and laws established by the legislative component of the regime, just as the Dispute Settlement Body (DSB) makes judgements on the basis of the WTO and the ICC interprets and applies the rules set by the Rome Statute. These courts could include a series of appeals, the provision of legal services and research, judges selected randomly, and would grant legal status to the relevant agents. What's more, this judicial body would be empowered to review the specific actions of member states to the regime. Member states would be able to bring suits against other members who are purported to act in violation of the treaty. Over time, these suits would generate a case law that would create and deepen a public and independent standard for evaluating subsequent state decisions. If appropriately structured, this global climate court can provide a mechanism for calling more powerful states to account and contesting decisions that will affect the less powerful in much the same way that domestic judicial proceedings can compensate for large differences in social or economic position within a state (Waldron 2011). Again, I want to remain agnostic about the particular structure of the court, but I do want to suggest that many international courts are inadequate because they grant standing to bring suit to too small a group of people. Allowing individual citizens, non-governmental organizations and independent ombudsmen to bring suits before the court would help avoid a common problem. Namely, states can be reluctant to bring suits because they worry about their relationship with the defendant in other contexts. Independent ombudsmen would help alleviate that problem.

Yet, this legislative-judicial regime fails. The legislative component ensures that the formulation of broad principles to regulate emissions behavior is not held hostage to the arbitrary deployment of power by unequal agents. The judicial component ensures that the application of these principles to particular, difficult cases will be done in a non-dominating fashion. Yet, there still remains a sense in which the regime relies on the power of states, and this will remain worrisome as long as states have radically different levels of power. Namely, the regime relies on states to *enforce* the laws and the judgements made by other components of the regime. And this is no small thing. Consider the example of President Andrew Jackson in the context of *Worcester v. Georgia.* The U.S. Supreme Court had set sharp restrictions on the legal right to remove Native Americans from their land in Georgia. Jackson, in response, was reported to have said, '[Chief Justice] John Marshall has made his decision; now let him enforce it!'[12] The Cherokee nation was eventually dispossessed and driven to a reservation. The take-away is this: U.S. law offered standing to a Native American tribe and the Supreme Court made a decision that protected their interests. Yet, this particular tribe was subject to the arbitrary power of the United States because the coercive enforcement of those laws and that judgement was dominating in virtue of the radical differences in power between the United States and the Cherokee. A similar objection can be made to our burgeoning environmental governance regime. While we have ensured that the regime will deliver judgements in a non-dominating fashion, we have not ensured that those judgements will non-dominatingly result in the appropriate deployments of coercive power to enforce the relevant political entitlements.

Consider the enforcement of the DSB (Lida 2004). In the event that the DSB rules that one state has engaged in unfair trade practices against another, the victim state is then *permitted* to engage in protective trade policies to punish the offending state. Yet, this creates a serious problem in the face of large disparities in power and influence. By relying on the victim state to sanction the offending state, the legal rights of the victim state are now subject to whether the victim state can effectively enforce those claims. So, in many cases, victim states that have been granted a legal right to sanction the offending state do not exercise this right, knowing that doing so could undermine their relationship with the more powerful state in other contexts. Perhaps the more powerful state will retaliate in a whole host of ways to attempts by the victim state to enforce its rights. Or more speculatively, we can imagine that many states will not even bring suits before the DSB because they know that, even if they succeed, they will not be in a position to sanction the defendant. These results are produced by a potent mixture of two features of DSB jurisprudence. First, the sanctioning relationship is dyadic: the plaintiff state is the only agent given the legal power to sanction the defendant state. Second, the sanction itself is contextualized to the regime: the plaintiff state can only sanction the defendant by deploying trade practices that would normally be protectionist under the treaty. This means that if the offending state is simply a lot less vulnerable to trade protectionism than the victim state, then the offending state can continue with their behaviour in the knowledge that they are in a superior bargaining position. In other words, even if we granted the dubious claim that existing WTO laws and DSB resolutions were designed such that they ensured equal standing, the enforcement apparatus would make it such that the actual enjoyment of WTO related entitlements would depend on differences in power and position. What's more, a natural way of solving the problem—granting a *universal* entitlement to impose sanctions in order to impose DSB resolutions to all states—would not prevent domination. The reason is that it would remain within the discretion of other states to determine whether to support the claims of the victim. And so, the rights of the victim would be dependent upon the decisions of other states over which they would have no control or ability to contest (Caney 2006, 746–47).

In order to respond to these worries, we should include an *executive* component of the regime that would be responsible for enforcing the legislative and judicial components through sanction. The executive could be empowered to legally obligate states to sanction other states through a variety of mechanisms: refusal to grant additional carbon permits, trade sanctions or embargos, and, in extreme cases, more coercive or forceful measures including sanctions and, perhaps, military force. One complication of environmental governance is that the contextual tit-for-tat WTO strategy is likely to be even less effective; the effects of a trade war are likely to be mostly felt between the participants. But the effects of an 'environmental emissions war' where states punish an emitting state by emitting themselves are likely to harm individuals unrelated to the dispute. Trade sanctions can be, at least somewhat, targeted at the appropriate offenders; emissions and emissions permissions cannot be. As a consequence, the sanctions will need to move beyond the legal entitlements—in, for example a cap and trade regime—that the governance regime distributes. So, the executive elements of our proposed regime will have a set of complex capabilities. First, it will contain a regime ombudsmen (Caney 2006, 746–47) that can bring suits against stakeholders who act against the regime in the judicial component. Second, it will need to possess some information-collection or monitoring capability in order to assist the 'emissions court.' And finally, the executive will make the determination as to how a particular judgement by the court will be enforced by deploying a variety of diplomatic, economic and political mechanisms that states will be collectively obligated to participate in and contribute to.

Our proposed global regime is well-placed to compensate for the large differences in power and vulnerability that currently drives the geopolitics of climate change governance. The legislative component ensures a set of emissions entitlements will be produced in a context where the equal standing of those subject to the regime are respected. The judicial component ensures that judgements concerning the application of those principles will be done in a way—both in terms of their constitution and procedure—that ensures genuine mechanisms of contestation and accountability obtain even between states that are unequal in capabilities. The executive component administers a sanctions regime that ensures that these judgements will not be differentially complied with based upon the calculus of power. What's more, the executive component will ensure the integrity of the regime by being empowered to bring suits itself on its own behalf. These three components work together. For example, the fact that sanctions will be distributed to compensate for differences in geopolitical power also gives smaller, less powerful agents at the legislative stage freedom to select principles without worrying about retaliation. Finally, there are mechanisms in place that allow for the non-dominating resolution of conflicts between the legitimate goals of this particular regime and other governance agents.

## THE NEW LEVIATHAN: A DOMINATING GOVERNANCE REGIME?

The regime described in the previous section can—at least in principle—be one that reduces or eliminates the specific kind of international domination related to climate change. Yet, in the process, we have created a governance regime of considerable power. After all, any regime that can implement sanctions that will compensate sufficiently for the power differences between the United States and Vanuatu will be need to be quite robust. But this gives rise to a different question: Does the new political agent itself dominate? After all, the decisions of the regime will significantly redistribute entitlements and, correspondingly, the life chances of particular people. In other words, there will be winners and losers in virtue of particular decisions made by our environmental regime, and those decisions will be backed by force. We have been focused on what the introduction of a third political agent does to the relations amongst the individuals of differentially powerful states, but what about the *new* relationship we will have been created between those who wield power on the basis of this regime? One might think that a similar question faced the early state-builders of the 16th-18th century. The creation of a centralized state had the effect of limiting arbitrary *private* violence—by eliminating feudalism and prohibiting private aristocratic armies—but at the cost of creating a new public authority that lacked any domestic rivals for power and which possessed unparalleled capabilities for violence. One can see the totalitarian regimes of the twentieth century as examples of this newfound state capacity being deployed to generate unprecedented levels of oppression. Furthermore, we can equally see the development of democratic politics, an independent judiciary, the rule of law, and other sorts of reform as efforts to ensure that this new power and authority were directed towards the prosperity and security of its subjects rather than the aggrandizement of those that happened to wield it. (See Tilly 1992 for a descriptive version of this story.) So, there is a danger of creating a regime—particularly in a context where considerable scientific expertise is required to understand and apply principles that distribute entitlements to emit—where technocratic expertise is used as an excuse to inoculate the actions of the regime from democratic oversight (Beckman 2008).[13]

One tempting way to prevent domination is through a kind of global, democratic politics (Held 1995). There seem to be two promising institutional manifestations of this kind of politics. First, we could create mechanisms by which individual subjects of the regimes could vote *directly* for potential magistrates. Second, individuals could vote for members of a representative body that would then make appointments to fill those positions. These institutional mechanisms would represent democratic oversight, generating a direct relation of accountability between the particular citizen and the regime. This should be combined with other elements of democratic rule: citizens being able to bring suits that can be appealed to in the global emissions court and being able to run for office themselves. This—combined with the rule of law, separation of powers, and checks and balances that are already elements of the regime—undermines or perhaps eliminates the arbitrariness of the power deployed by the regime.

This strategy, though possessing a certain amount of appeal, has some drawbacks. Namely, it places a high cognitive load on individual citizens while stretching the bounds of common affection or identity that might be required for democratic politics. This seems especially true because of the nature of the climate impacts the regime purports to regulate. The effects of the emissions behaviour of individuals are global and the empirical knowledge needed to judge policy would be very high. The problem here is not simply that climate change is comparatively difficult to understand in virtue of its global nature and its scientific complexity. It is also that it touches on cultural matters concerning what kind of life is worth living and one's relationship to nature that are difficult to learn and fully comprehend. And one will need to understand how these concerns will translate into unfamiliar political idioms, narratives, and lexicography. It is a lot to demand. What's more, the feedback loop generated by democratic politics is considerably more attenuated in the case of climate change. (Sen 1999 describes the feedback loop.) Particular individuals—in the context of economic failures or widespread human rights violations—are likely, if there are appropriate democratic institutions, to be able to aggregate that information and respond to it in effective ways. Yet, the temporally and spatially diffuse nature of the effects of climate change makes the normal operation of democratic politics difficult. To fully grapple with climate change in the context where demos and affected agents completely overlap, individuals would need to understand themselves as part of a global demos that extends far into the future. This is difficult both in terms of individual imagination and in terms of structuring democratic institutions. We are, quite frankly, both personally and politically all too likely to be biased in favor of the near and dear, both spatially and temporally. And it is almost certainly true that without dealing with the intergenerational issue, even fairly effective global governance will face serious difficulties in adequately responding to climate change (Parfit 1984; Gardiner 2011 Part C; Beckman 2008; Smith 2013). Furthermore, the global nature of the policy problem means that democratic politics will need to operate across a wide variety of communities, operating with different languages, idioms, and cultural understandings. This, presumably, will make it difficult for individuals to view themselves as engaged in a collective project and to reliably impose electoral costs when the institutions fail others with whom they fail to share the relevant cultural substratum (Kymlicka 2001).

Most of these worries about global democracy are familiar, but my analysis adds two new elements. First, the nature of climate change, its impacts and its potential policies uniquely exacerbate these problems. Even in fairly complicated cases—such as trade—the interactions are comparatively discrete and dyadic. That is, trade regulations cover global interactions—contracts, the movement of goods and services and the like—that generally affect members of a limited number of states. Furthermore, the *effects* of those interactions primarily occur within the states whose citizens are participating in the interaction and these effects can be attributed to a discrete set of interactions. In climate change, this is less true and, as a consequence, regulation needs to be more holistic and global in orientation. Second, domination offers a new reason *why* these worries are problematic. Namely, democratic politics does not only serve an aggregative or epistemic function (Anderson 2006) but also places an important external check on the operation of political institutions. This point operates in two directions. First, the cognitive requirements for a global, environmental politics and the lack of a cultural baseline are merely *contingently* related to its potential failure as a strategy for non-domination. I am not claiming that democracy is necessarily connected to a cultural demos and that this makes global democracy impossible. Nor, conversely, would a shared culture necessarily resolve the relevant problem; what matters is whether individuals have a reasonable opportunity to act collectively to impose real, material costs on the exercise of political power. Second, merely demonstrating that a global democracy—however imperfect—would serve an epistemic function would be insufficient, and this makes the case for global democracy much more difficult. Active decision-making, and not mere aggregation of preferences, is what is needed. So if the value of global democracy depends entirely on global governance agents virtuously taking the judgements of the democracy seriously, then we have *not* solved the problem of global domination simply because those agents tend to listen to democratic institutions when they speak.

There is a final worry about global democracy that depends less on the way in which a global democracy might function and more on the kinds of relations that domestic democracies make possible. Let's grant, for the sake of the argument, that at least some of the states that will be subject to the regime are themselves domestically non-dominating. These relations are sufficiently accountable that the individuals who constitute them are adequately free. Of course, this does not mean that these states are adequately just: they dominate outsiders and foreigners by, for example, engaging in unilateral emissions behaviour. However, these non-dominating internal relations may offer a reason to be somewhat skeptical of attempts to resolve international domination that involve disrupting them. After all, to do so would be risky, we would be revising or reforming institutions that successfully make free relations possible with the hope that the new institutions we create build on those successes and create free relations in the contexts in which the previous institutions fell short. It seems plausible that the justification of this sort of disruptive reform will require that this risk be taken into account. So, if we think that there are independent reasons for skepticism about global democracy, then a presumption against disrupting political institutions that create at least *some* non-dominating relations may remain undefeated (Kleingeld 2011). This argument is parasitic on whatever skepticism one might have about the restraining effects of global democracy. The claim is that—if there are other options—we need *comparatively* weaker reasons to adopt these other non-dominating strategies even if they are comparatively more expensive in terms of other values, that do not risk domestically non-dominating constitutional orders. For example, we might need to countenance less economically efficient regimes—or regimes that are less effective at protecting ecological values—if those regimes instantiate a lesser risk to non-dominating domestic orders.

These problems imply a second strategy of non-domination: an institutional division of labor. States can intervene between the regime and particular persons. If those states serve as effective sites of collective, political organization, then they can appropriately constrain the potentially dominating force of the environmental regime. In other words, states can *mediate* between the regime and its subjects, offering an independent venue for accountability and contestation. The institutional setup would be indirect: citizens would participate directly in the election of their own state officials and those state officials would then participate in selecting candidates who would take up the offices that compose the environmental regime. Thus, individual citizens would have a powerful advocate for their interests that they could hold accountable who could, in turn, effectively participate in the governance of the broader regime. If the regime itself was constructed such that state participation was structured in ways that compensated for differences in power and wealth, then we could allow states to serve as the sites of democratic politics while, at the same time preventing more powerful states from dominating the proceedings of the regime (Kymlicka 2001).

There are two problems with this proposal. First, this mechanism of accountability only resolves the problem of domination *if the state is an effective and adequate advocate for the interests of its citizens*. In other words, states will need to be constructed internally so that they actually are sites of contestation and accountability. If states dominate their own citizens or do not adequately protect them from the private domination of others, then their participation in the regime will simply be another locus of domination. Or rather, the regime itself will dominate those citizens insofar as it deploys power over them, and this will be true regardless of whether the regime deploys that power in ways that serves the interests of those citizens. So, while it may be true that a mediated regime that was indirectly accountable to individuals via their own states could be non-dominating, minimally accountable and contestable institutions at the state level would be necessary. As a consequence, global justice cannot be separated from questions of climate change justice. If the international system and its component states are not sufficiently accountable to the fundamental interests of their members, then the environmental regime will not be able to exercise power rightfully. This is a serious problem because many states are not adequately accountable to their citizenry.

The second problem is that it is hard to imagine various ways of granting independent authority to states in order to serve as a check on the power of the environmental governance regime that would not reduce the effectiveness of that regime. That is, it seems plausible that we might be able to create legislative and judicial mechanisms that can mostly compensate for differences in power and wealth. After all, such compensation seems possible domestically. Yet, some state interests *will not align* with what we would take to be the ideally just distribution of entitlements to emissions and adaptation compensation. So, if a state—through perfectly democratic procedures—rejects the sacrifices being demanded of it, either because the regime is too demanding or insufficiently demanding, then the regime will simply force the state to accept to sacrifice or the regime will grant the state discretion to make those sorts of decisions. If the latter, then the regime is unlikely to be effective as states that are less vulnerable will exercise that discretion in ways that serve their interests. Granting states independent authority to criticize and abstain from the regime will undermine the regime's effectiveness. In other words, granting state discretion a large role in preventing domination will re-instate the superior bargaining position of emitters that undermines our current efforts to react more robustly to climate change.

## CONCLUSION

A powerful global governance regime is needed to redistribute and reform coercively enforceable entitlements concerning the emissions of greenhouse gases in the face of large scale differences in power and vulnerability between states. This regime will need to have legislative, judicial and executive components that will generate a serious risk that it will itself be dominating. There are two broad strategies for making this regime accountable to its members: direct and indirect. Direct mechanisms allow individuals to directly participate in the regime by bringing suits, running for office and electing officials. Indirect mechanisms rely on the state as a nexus of democratic decision-making for individuals and treats the state as the unit of political participation at the level of the regime. Yet, both types generate serious worries. Direct mechanisms might simply be too demanding, forcing individuals to shoulder unsustainable cognitive and motivational loads and consequently risk the disruption of non-dominating domestic orders for little reward. Indirect mechanisms, on the other hand, rely almost entirely on the domestic justice of states to generate the relevant accountability and so cannot be used for states that are not minimally legitimate. What's more, by granting more power to states, the regime may be readmitting the large differences in the interests of rich and poor countries into policy-making concerning climate change. As a consequence, indirect mechanisms are more likely to undermine the *effectiveness* of the regime. This gives rise to a serious dilemma: a regime that relies on direct mechanisms will likely be effective but dominating. The reason is that the cognitive and motivational loads will undermine the capacity of citizens to engage in the collective political actions that are needed to hold the regime to account. While a regime that relies on indirect mechanisms will likely lead to more coordinate political action that will hold the regime to account, the cost of doing will be to grant discretionary powers to states that—while now non-dominating—will likely lead to a less effective emissions regime as that discretion will tend to be exercised in the name of the specific interests of each state's constituents. If each state was situated in a symmetrical fashion with regards to climate change impacts, then this might not be a problem. Unfortunately, that is not true. As a consequence, we will likely face a tradeoff between effectiveness and domination.

It is a difficult question how to balance these considerations, but I am tempted to favor the direct strategy. The reason is that while—in principle—we could imagine a global order that compensated for imbalances in the geopolitical power in the formulation of an emissions regime that granted sufficient discretion to individual states in order to be non-dominating, it would be difficult for that system of discretion to avoid recapitulating the power relations that drove the creation of the regime in the first place. In other words, the indirect strategy—like many federal regimes—is unstable; too much discretion and we return to *international* domination while too little discretion results in *global* domination. The direct strategy, by contrast, is theoretically more stable: democratic politics plays the same functional role in the global context as it does in the domestic; it is simply scaled up. The problem with the direct strategy is that we will need to take risks—breaking up domestically just relations—with the hope that, as we do so, the motivational and cognitive limitations will be overcome.

None of this implies that we should not continue with the UNFCCC process and attempt to limit, as far as we can, the negative impacts of climate change. Perhaps it is true that any reasonable attempt to create a non-dominating political order will require that we take the initial steps we are now taking. Yet, there are still several reasons why it matters that our current order dominates and will continue to do so for the foreseeable future. First, there may come a time where we are in the position to sacrifice a particular value—economic growth, for example—in order to create a non-dominating order and we need to be aware of those opportunities. Second, some agents may be owed compensation in virtue of the fact that they are being dominated and this may influence the distribution of emissions entitlements in the long run. Finally, certain actions that would normally be impermissible—geoengineering or the cancelling of foreign debt—may be permissible in the context where one is being dominated. In other words, knowing whether we dominate may direct our potential efforts at reform, justify new claims to compensation and undergird claims to engage in revolutionary action in order to change the current system. Domination, then, leaves a significant moral remainder in our theorizing about emissions governance even if it has no immediate effect on our current policy.

## REFERENCES


Anderson, Elizabeth. 2006. 'The Epistemology of Democracy'. *Episteme* 3: 8–22.

Beckman, Ludvig. 2008. 'Do Global Climate Change and the Interest of Future Generations Have Implications for Democracy'? *Environmental Politics* 17: 610–24.

Blomfield, Megan. 2013. 'Global Common Resources and the Just Distribution of Emission Shares'. *Journal of Political Philosophy* 21: 283–304.

Broome, John. 2006. *Climate Matters.* New York: W.W. Norton and Company.

Bull, Hedley. 1977. *The Anarchical Society*. London: Palgrave McMillan.

Caney, Simon. 2006. 'Cosmopolitan Justice and Institutional Design'. *Social Theory and Practice* 32: 725–56.

———. 2012. 'Just Emissions'. *Philosophy and Public Affairs* 40: 255–300.

Drezner, Daniel. 2007. *All Politics Is Global*. Princeton, NJ: Princeton University Press.

Dryzeck, John. 2005. *The Politics of the Earth*. Oxford: Oxford University Press.

Filippov, Mikhail, Peter Ordeshook, and Olga Shvetsova. 2004. *Designing Federalism: A Theory of Self-Sustainable Federal Institutions*. Cambridge: Cambridge University Press.

Firger, Daniel, and Michael Gerrard. 2011. 'Climate Change and the WTO: Expected Battlegrounds, Surprising Battles'. *Daily Environment Report of the Bureau of National Affairs* 2011.

Gardiner, Stephen. 2002. 'The Real Tragedy of the Commons'. *Philosophy and Public Affairs* 30: 387–416.

Gilpin, Robert. 1981. *War and Change in World Politics.* Cambridge: Cambridge University Press.

Gosseries, Axel. 2011. 'On Future Generations' Future Rights'. *The Journal of Political Philosophy* 16, no.4 (2011): 446–74.

Jameison, Dale. 2005. 'Adaptation, Mitigation, and Justice'. *Perspectives on Climate Change*: *Science, Economics, Politics, and Ethics*, Volume 5: 217–48.

Hart, H. L. A. 1961. *The Concept of Law*. Oxford, Oxford University Press.

Held, David. 1995. *Democracy and the Global Order*. Stanford: Stanford University Press.

Kant, Immanuel. 1996. *The Metaphysics of Morals*, translated by Mary Gregor. Cambridge: Cambridge University Press.
Kleingeld, Pauline. 2012. *Kant and Cosmopolitanism: The Philosophical Ideal of World Citizenship*. Cambridge: Cambridge University Press.
Kymlicka, Will. 2001. *Politics in the Vernacular.* Oxford: Oxford University Press.
Laborde, Cecile. 2010. 'Republicanism and Global Justice: A Sketch'. *European Journal of Political Theory* 9: 48–69.
Lida, Keisuke. 2004. 'Is WTO Dispute Settlement Effective?' *Global Governance* 10: 207–25.
Lovett, Frank. 2012. *A General Theory of Domination and Justice.* Oxford: Oxford University Press.
Neumayer, Eric. 2000. 'In Defence of Historical Accountability for Greenhouse Gas Emissions'. *Ecological Economics* 33: 185–92.
Parfit, Derek. 1984. *Reasons and Persons*. Oxford: Oxford University Press
Pettit, Phillip. 1997. *Republicanism: A Theory of Freedom and Government*. Oxford: Oxford University Press.
———.2010. 'Legitimate International Institutions: A Neo-Republican Perspective', in *The Philosophy of International Law,* eds. Samantha Besson and John Tasioulas. Oxford: Oxford University Press.
———. 2013. *On the People's Terms: A Republican Theory and Model of Democracy*. Cambridge: Cambridge University Press.
Pogge, Thomas. 2002. *World Poverty and Human Rights*. Polity: Cambridge.
Posner, Eric, and David Weisbach. 2013. *Climate Change Justice*. Princeton, NJ: Princeton University Press.
Ripstein, Arthur. 2004. 'Authority and Coercion'. *Philosophy and Public Affairs* 32: 2–35.
———. 2009. *Force and Freedom: Essays in Kant's Political and Legal Philosophy*. Cambridge: Harvard University Press.
Roughan, Nicole. 2013. *Authorities: Conflicts, Cooperation, and Transnational Legal Theory*. Oxford: Oxford University Press.
Sen, Amartya. 1999. *Development as Freedom*. New York: Alfred Knopf.
Smith, Patrick Taylor. 2013. 'Domination and the Ethics of Solar Radiation Management', in *Engineering the Climate: The Ethics of Solar Radiation Management*, ed. Christopher Preston. Lanham, MD: Rowman & Littlefield.
Stilz, Anna. 2009. *Liberal Loyalty*. Princeton, NJ: Princeton University Press.
Tilly, Charles. 1992. *Coercion, Capital, and European States: AD 990–1992.* Malden, MA: Blackwell Publishers.
Vanderheiden, Steve. 2008. *Atmospheric Justice: A Political Theory of Climate Change*. New York: Oxford University Press.
Waldron, Jeremy. 2011. 'How Law Protects Dignity'. *NYU School of Law, Public Research Paper No. 11-83*. http://dx.doi.org/10.2139/ssrn.1973341
*Worcester vs Georgia*, 31 U.S. 6 Pet. 515 515 (1832)
Young, Iris Marion. 2006. 'Responsibility and Global Justice: A Social Connection Model'. *Social Philosophy and Policy* 23: 102–30.
IPCC, 2014: Climate Change 2014: Impacts, Adaptation, and Vulnerability. Part A: Global and Sectoral Aspects. Contribution of Working Group II to the Fifth Assessment Report of the Intergovernmental Panel on Climate Change [Field, C.B., V.R. Barros, D.J. Dokken, K.J. Mach, M.D. Mastrandrea, T.E. Bilir, M. Chatterjee, K.L. Ebi, Y.O. Estrada, R.C. Genova, B. Girma, E.S. Kissel, A.N. Levy, S. MacCracken, P.R. Mastrandrea, and L.L. White (eds.)]. Cambridge University Press, Cambridge, United Kingdom and New York, NY, USA, 1132 pp.

## NOTES

1. Center for Ethics in Society, Stanford University. I would like to thank the members of the Center for Ethics Postdoctoral Workshop for all of their help. Aaron Maltais and Catriona

McKinnon provided extensive comments and showed enormous patience in making the chapter better. I cannot thank them enough. The mistakes are mine.

2. See Megan Blomfield's contribution to this volume for a more detailed explication of these effects.

3. One possibility is that the new political system will *return* some discretionary power to current states. I explore that possibility later. Yet, there will be a sense—if a non-dominating emissions order is created—that states are no longer sovereign and final decision-makers even if that is so.

4. Of course, any significant socio-political change will make some actions more costly or impossible while making other possibilities less costly or newly possible. The insight of the domination account is that we cannot understand whether someone is free by simply looking at what options they happen to have. Rather, we must understand how they came to *have* the options they have and whether those options were constrained in legitimate ways.

5. This claim does not commit us to a problematic 'explanatory nationalism' (Pogge 2002) by which the international system plays a very marginal role in state capacity and prosperity. The international system may help or hinder development of some nations, but the international system has primarily been—for most of its history—a 'society of states' that depends on the already powerful for its maintenance and enforcement (see Gilpin 1981 and Bull 1977). Powerful states might be manipulating the international system to favor their interests, but a large part of the reason these states are powerful is because of their domestic economies, technological superiority and military capacity. Their unjust dominance of the international system is generated by their power; it does not easily explain it.

6. This is a serious worry for Pettit's (2010) solution to international domination. See Philip Pettit 'Legitimate International Institutions'. In the context of environmental governance, it is very likely that rich, emitting countries will band together to impose a regime on weak, vulnerable countries. The reason is that, unlike trade or security, the convergence between intrinsic power and the interest in emitting is very strong; this makes it hard to imagine a balancing coalition developing.

7. Some of these impacts will be, as Laborde (2010) has described, 'capacity-denying' acts of domination whereby individuals and states will be less able to protect themselves and others from domination. However, even if those impacts did not reduce *those* particular capacities, the structuring of their life chances would still be an instance of domination.

8. For a description of the accord, see: http://www.nytimes.com/2014/11/12/world/asia/china-us-xi-obama-apec.html?_r=0

9. If the global order non-dominatingly distributed an entitlement to each state such that they could determine—with complete discretion—how much to emit, then there would be no problem of international domination. That distribution of entitlements would be immensely imprudent and there would remain the problem of intergenerational domination.

10. This is, more or less, the conception of the UNFCC.

11. To put it another way, if we thought that an important reason to favor a global governance regime was that it represented a pre-commitment strategy to ensure compliance to the collective self-interest in the face of incentives to defect in a tragedy of the commons, then having each state judge cases for themselves will simply recapitulate the very same problems that motivated the pre-commitment strategy in the first place.

12. This example is a bit more complicated than is normally recognized because the Supreme Court did not order the President to enforce an order as result of the case. As a consequence, it was left to the discretion of the President to enforce the law as interpreted by the court. This detail actually makes the case more applicable to my point; the Supreme Court was allowed to grant discretion to the executive in the enforcement of the law.

13. Dryzeck (2005) makes a similar argument, especially in Chapter Four.

# *II*

# Democratic Legitimacy and Equity in Climate Governance

*Chapter Three*

# Climate Change Duties and the Human Right to Democracy[1]

Ludvig Beckman

There is growing consensus that present and future effects of anthropogenic climate change constitute a real threat to human rights. The complex and far-reaching consequences of climate change include, but are not limited to, the loss of inhabitable and arable land, water shortages and damages to life and physical structures, due to rising sea levels, flooding and extreme weather, which in the end contribute to undermine human rights to life, health and subsistence (Bell 2011; Caney 2010). Though 'human security' is the preferred lingua of the 2014 IPCC assessment report, numerous human rights bodies acknowledge that climate change puts human rights at risk (McInerney-Lankford et al. 2009, 8; IPCC 2014, chap. 12).[2] Human rights mean to identify what governments 'must' do, not just what they 'should' do (Bodansky 2010, 517). Human rights law is consequently thought to be binding primarily on states as it obliges them to 'respect, protect, and fulfill' the entitlements of individuals (Green 2001, 1071).

Yet, human rights reflect claims about morally required rights and duties (Beitz 2001; Nickel 2010). The language of human rights serves to remind us that the most urgent interests of mankind are at stake, creating moral duties to secure the 'line beneath which no one is to be allowed to sink' (Shue 1996, 18).[3] Once our attention is shifted from legal to moral duties, the presumption that human rights are binding only on existing governments within the current state-system appears too limited.[4] The human rights tradition includes the notion that duties to respect or protect human rights apply to individuals no less than to governments and organizations (Gilalbert 2011). The idea of individual duties may seem impractical given that protection of human rights arguably depends more on institutions and policies than on what individuals taken separately are doing (Nihlén Fahlqvist 2009). But this is exactly why the duties applicable to individuals should be understood as demands to support and create the political institutions required for their protection. As argued by Ashford (2006), the foremost duty of individuals in relation to human rights is to push for institutional reform that allocates duties and responsibilities in ways that are effective to their protection. In the context of climate change, Baylor Johnson (2003, 272) similarly defends climate duties to 'work for and adhere to a collective scheme' equipped with the required capacities.

Now, many believe that political institutions characterized by electoral democracy at the national level are particularly ill-designed to respond to the challenge of climate change (Gardiner 2013, 227; Gardiner 2006). The nature of the 'collective scheme' and the extent of 'institutional reform' required to protect people from climate change might in other words diverge from the institutions of electoral democracy. Duties to protect human rights from the consequences of climate change are perhaps best understood in line with Derek Bell's account: 'Each of us has a general duty to promote effective institutions for the specification, allocation and enforcement of more specific duties that will provide a significant level of protection for current and future persons' basic rights from the effects of anthropogenic climate change' (Bell 2011, 119).

It is worth paying attention to what is entailed by Bell's claim. As it stands, it holds that a sufficient reason for the existence of duties to promote an institutional framework is that it possesses the capacity to protect human rights from the harmful effects of climate change. That is, the object of our institutional duties is determined exclusively on the basis of the outcomes that the institutions are expected to produce. This claim is radical as it does seem to imply that the procedural qualities of institutions are relegated to secondary importance. Hence, Bell's account does not attach any importance to many of the defining components of electoral democracy, such as universal suffrage and executive accountability.[5]

The point here is that climate duties grounded in a concern for human rights raise fundamental challenges to anyone who believes that democratic institutions generate claims of allegiance, encapsulated by duties to support democratic procedures. The challenge then is how to deal with the potential conflict between duties to promote institutions with the capacity to protect human rights from climate change, and duties to support and respect the workings of democratic procedures. The purpose of this chapter is to take some preliminary steps towards teasing out the conditions for when the conflict emerges, thereby contributing to the clarification of the relationship between climate duties and democracy.

When the individual is subject to conflicting duties, there is usually some way to resolve the conflict by systematically thinking through the relative strength of the considerations that ground the duties in the first place. In so doing, we may be able to conclude that one set of duties should take precedence before another. On the other hand, if the clash of duties depends on the fact that there are conflicting rights-claims, it appears that there is further question to be examined, namely, to what extent their rights are really compatible.[6] If they are not, we may conclude either that there is genuine conflict of rights or that what appeared to be the rights of some people should not be so conceived at all. In the latter case, the situation is more appropriately characterized as a pseudo-conflict where on inspection it is clear that one of the competing claims does not represent a justified rights-claim (Griffin 2008, 68; Kamm 2001). In the former case, where genuine rights-claims compete, it follows that respect for rights is inconsistent with respect for every rights-claim.[7]

These distinct questions bear some similarity to the distinction between the agent-relative and the agent-neutral approaches to rights (Sen, 1982; Kamm 2001; Tasioulas 2010). The former considers how individuals should resolve competing moral demands in a given situation. The conflict basically takes place within the agent. By contrast, the agent-neutral viewpoint considers conflicts between the rights of different people irrespective of the demands placed on agents in specific situations. From the agent-neutral perspective, the conflict does not reside within a single duty-bearer but between several rights-bearers.

In what follows we examine the suggested conflict between climate duties grounded in respect for human rights and duties to honor democratic institutions first from the perspective of the individual duty-bearer and, only secondly, in terms of conflicting rights-claims. The first and second sections engage the issue from the agent-relative perspective. Here, we explore whether the conflict can be avoided by separating different types of duties and, secondly, whether the conflicting interests can be balanced. Only when we have explored these attempts of reconciliation between climate duties and democratic duties do we engage the potential conflict between rights. The final section defends the view that the conflict between democratic rights and human rights applies only intergenerationally.

## DEMOCRATIC DUTIES AND CLIMATE DUTIES

Duties to protect human rights from climate change impose demands on us that run counter to a range of other ends. If we allow for the existence of climate duties, we should expect that they incur some costs in the sense that they entail demands overriding otherwise morally permissible or required aims. Climate duties are presumed to demand that individuals do more than they currently are doing (Raterman 2012). Following this attempt to characterize conflicts between climate duties and other moral requirements, it is doubtful that such situations represent moral conflicts in any genuine sense at all. Climate duties are overriding and that is all there is to it. A second approach takes the notion of conflict more seriously and argues that when different duties conflict that is typically because the duties involved are all overriding. When these overriding duties conflict, the nature of the dilemma is more difficult to pin down. One suggestion is that we are then facing a 'strong moral dilemma', meaning that the individual is subject to moral requirements that conflict because they are overriding. Yet, to characterize any situation in those terms appears incoherent. Thus, in case two or when more moral requirements conflict, it must be true either that one of them is overriding or that none of them are (Sinnott-Armstrong, 1996, 50).

Yet another suggestion is that conflicts between duties signal uncertainty about the nature of our moral requirements. The claim that climate duties conflict with duties to support the institutions of electoral democracy thus only serves to report the unfinished state of our understanding of their relationship. The conflict obtains prima facie but not all things considered (Brink, 1994).

Following the third approach, the relevant task is to spell out in greater detail the various reasons that determine our duties all things considered. Consider, to begin, the nature of duties to support the institutions of electoral democracy. Various justifications for the existence of such duties can be developed. Following Rawls, each member has a 'natural duty' to support political institutions that comply with the demands of justice (Rawls, 1971, 115). Thomas Christiano argues that citizens have duties to 'comply with democratic decisions' if they are consistent with the moral requirements of 'public equality' (Christiano 2009). Moreover, David Estlund defends the authority of collective decisions authorized by democratic procedures, even if erroneous, by reference to the overall epistemic benefits of compliance with such procedures (Estlund 2009).

Though the grounds for democratic duties are varied, I will bypass the differences in order to focus on what they have in common. In what follows, I will use 'democratic duties' as short for any account that considers citizens to be subject to (prima facie) duties to comply with or support democratic decisions where such decisions are defined primarily in procedural terms. The procedural component excludes the possibility of defining democratic decisions in terms that would make them coextensive with specific policy outcomes, such as would presumably be stipulated by climate justice and its correlative climate duties. That is, the procedural basis for democratic duties is specific enough to distinguish them from climate duties as the latter is by definition concerned only with the substantive outcomes of collective decisions.

Democratic duties include a variety of distinct moral requirements, some negative and others positive in character. They might for instance entail duties of assistance, directed towards the maintenance of democratic institutions, and duties of non-interference grounded in respect for decisions and policies generated by institutions that are democratic.[8] To the extent democratic duties apply, they seem to negate any account of duties that does not confirm duties to support democratic procedures. Hence, the claim that citizens do have climate duties that require support for the creation of institutions with the capacity to protect human rights from climate change potentially conflicts with the claim that citizens have democratic duties. The conflict represents a dilemma for the individual citizen in so far as he or she is unable both to support democratic institutions and to support the creation of a completely different set of political institutions dedicated to the eradication of dangerous climate change.

Of course, the general claim that procedural democracy generates democratic duties when they correspond to the ideals of justice, public equality or epistemic benefits does not automatically vindicate the more specific claim that the citizens of electoral democracy have duties to support *those* institutions. Electoral democracy is a concept referring to certain minimum requirements of democracy, including regular elections free from large scale fraud under conditions of universal suffrage, and so on (Munck 2009). Yet, the fact that some existing political institutions meet the criteria of electoral democracy does not prove that they satisfy the normative criteria of justice, public equality or epistemic democracy. Hence, if democratic duties are premised on certain normative criteria, it would remain an open question to what extent democratic duties apply to citizens of electoral democracy. Since it seems somewhat unlikely that electoral democracy fully satisfies these normative criteria, democratic duties are unlikely to apply to citizens of such regimes. The suggestion that citizens are torn between climate duties and democratic duties consequently seem to be without foundation.

The argument that the conflict is unlikely to apply can be made also in terms of the contrast between duties applicable in ideal conditions and duties applicable in non-ideal conditions. Democratic duties evidently apply in ideal conditions, where general compliance with the moral requirements of democracy and justice obtain. By contrast, duties to promote institutions protecting human rights from climate change seem applicable in non-ideal conditions, where full compliance with the demands of justice does not obtain.[9] Now, there can be no conflict between duties applicable in imperfect conditions and duties applicable in perfect conditions, and so the presumed tension between climate duties and democratic duties would be nothing but a pseudo-conflict.

Now, the objection that democratic duties apply only in ideal conditions and therefore fail to be relevant in electoral democracy is not fully convincing. Instead, there is reason to insist that principles of justice and public equality generate individual duties to support political institutions also under non-ideal conditions. The rationale for this view is that citizens have duties to pursue the realization of public equality and justice even in circumstances where political institutions do not currently honor its demands. In non-ideal conditions, democratic duties require that citizens support institutional reform. The image of a conflict between democratic duties and climate duties may consequently re-emerge, albeit slightly reformulated. On the one hand, individuals are presumed to have duties to support the creation of institutions with the capacity to protect human rights from climate change. On the other hand, individuals are presumed to have duties to support reform of existing political institutions in order to perfect democracy and justice.

Responding to this scenario, advocates of climate duties might contend that duties to support institutions preventing violations of human rights should take precedence. They should do so because climate duties derive from human rights and human rights should be granted 'priority' before other moral demands (Caney 2010). The basis for this premise is that human rights are the means for the preservation of people's most basic interests; they are 'minimum standards' for adequate life (Buchanan 2005; Nickel 2010).[10] The conjunction of the priority thesis and the claim that individuals have duties to promote institutions with the capacity to preserve human rights against the threat of dangerous climate change consequently offer powerful reasons against attending to duties to support democracy and justice within electoral democracies. In this spirit, some defend the view that climate duties justify 'non-violent civil disobedience' by citizens in democracies (Lemons & Brown 2011), whereas others endorse the even more radical view that citizens have duties to support the creation of authoritarian forms of government (Shearman & Smith 2007). There is consequently no claim-dilemma, as climate duties must be granted priority in every instance.

The view that climate duties should be granted priority because they are human rights duties is not conclusive in every instance, however. In order to see why, we need to spell out in greater detail the implications of the priority thesis. The claim that human rights have priority because they are grounded in the fundamental importance of certain interests of the individual[11] equals the claim that duties corresponding to human rights should be granted priority before duties associated with the preservation of other interests. Yet, any plausible theory of rights must acknowledge that they are associated with a range of distinct duties, some that are negative in the sense of requiring non-interference with the relevant interests, others that are positive in the sense of demanding preventive action, assistance or compensation, towards the rights-bearer (Shue 1996; Waldron 1989). The claim that some specific right should be granted priority before another specific right consequently translates to the claim that some duties should be given priority before other duties. But here is the crux. While the priority of human rights by definition entails that *some* duties required by respect for human rights should be granted priority before some duties required by respect for rights that are not human rights, it does not follow that *every* duty associated with human rights have priority before all duties associated with rights that are not human rights. Indeed, a moment of reflection is enough to indicate that there are at least some duties associated with human rights that do not have priority before duties associated with other rights.

If true, as argued by Pogge (2009), that negative duties are generally stronger than positive duties, it appears conceivable that some positive duties associated with human rights should not be granted priority before the negative duties associated with other kinds of rights. That is, just because reasons for not interfering with human rights are generally stronger than reasons for not interfering with other rights (because human rights have priority) it does not follow that reasons for offering assistance to the protection of human rights (positive duties) are always stronger than reasons for not interfering with other rights (negative duties).[12] Thus, the priority thesis is consistent with the claim that democratic duties include negative duties that are more important than at least some positive duties derived from human rights. For example, it may be the case that citizens have overriding reasons to comply with democratic duties not to interfere with democratic decisions even if citizens also have positive duties to support the creation of institutions with the capacity to protect human rights. Or, rather, in cases where positive human rights duties run counter to negative democratic duties, the priority of human rights does not determine how the conflict ought to be adjudicated.

The priority thesis remains decisive, nevertheless, in cases where negative climate duties conflict with negative democratic duties and, also, in cases where positive climate duties conflict with positive democratic duties. That is, whenever citizens face conflicts that are symmetric in the sense of being constituted by duties of the same type, the priority thesis entails that there is no genuine moral conflict at all since it should be clear which duties are overriding. In such cases, climate duties ought to prevail due to the superior moral significance of the interests they serve to protect. In addition, there may be situations where negative climate duties run counter to positive democratic duties. The outcome here should be equally evident, both because negative duties take precedence before positive duties and from the fact that human rights have priority before other rights. The conclusion, then, is that the priority of human rights helps reduce the range of conflicts between climate duties and democratic duties. The only genuine conflict that remains is where positive climate duties run counter to negative democratic duties. In such cases and only in such cases does it remain unclear which duties are overriding.

Now, the ability to reduce the conflict to such a large extent is premised on the claim that the priority thesis does not apply to democratic duties. The duties to comply with or to support democratic institutions are in most cases overridden by climate duties since only the latter are associated with the preservation of human rights. But it is worth contemplating the possibility that the premise of this inference is mistaken in the sense that we are in effect confronted with a conflict between human rights on both ends. If that is the case, the view that climate duties should be granted priority because *only they* serve to protect human rights would be unwarranted. Instead of a conflict between duties to protect human rights and duties to protect democratic institutions, we are facing a conflict between two different sets of duties that are both grounded in a concern with human rights.

## CONFLICTING HUMAN RIGHTS DUTIES

In order for democratic duties to be human rights duties, there must be human rights to democracy. The existence of such rights in the relevant sources of international law is presently uncertain.[13] Yet, as we are concerned with conflicts between moral duties, the relevant question is rather if human rights to democracy can be morally justified. An influential argument to the effect that we ought to recognize such rights is grounded in evidence of a positive empirical relationship between democratic institutions and the preservation of human rights. Following Charles Beitz (2001, 278) democracy should be included in the list of morally required human rights on condition that democratic rights serve as 'means' to 'the satisfaction of urgent human interests'. A similar position is taken by Thomas Christiano (2011, 144) arguing that human rights to democracy are justified because such institutions have proven to be 'necessary and reliable arrangements' for the protection of 'urgent moral goods'. Given present conditions of human affairs 'human rights require democracy', argues James Griffin (2008, 254).[14]

Evidence in support of the instrumental justification of a human right to democracy derives from observations about the empirical relationship between type of governments and the urgent interests of citizens. The test is whether citizens in democratically ruled nations enjoy significantly better protection of their human rights than citizens in non-democratic nations. Citing a wealth of empirical work that compares the fulfillment of human rights in democracies and non-democracies, numerous studies confirm that democracy provides a 'reliable arrangement' to the preservation of human rights.[15]

If these considerations are conclusive to the justification of human rights to democracy, they lend increasing force to democratic duties.[16] The duties of citizens to support democratic institutions and to comply with its decisions are no longer grounded in the interests of citizens qua citizens in participatory forms of government but are grounded in the interests of citizens as human beings in the preservation of basic human rights. That is, democratic duties are required by respect for human rights.

On the basis of this understanding, the conflict between democratic duties and climate duties needs to be reexamined. When the duties involved are both derivates of human rights, the notion that priority should be given to human rights is no longer helpful for the purpose of mitigating tensions between them. The conflict, if indeed there is one, is now located within the framework of human rights itself and is perhaps best summarized as the tension between duties to preserve human rights from degradation by the consequences of climate change and duties to preserve human rights from degradation by authoritarian governments. On the one hand, climate change puts human rights at risk, generating climate duties in order to alleviate the threat. On the other hand, human rights are put at risk by authoritarian governments, triggering duties to protect and support democratic forms of government. Hence, the conflict must be decided either by determining the relative importance of the interests involved or by questioning the assumption that the one set of duties cannot be consistently discharged without compromising the other set.

## Normative Solutions

The first option is to balance the interests threatened by authoritarian regimes against the interests threatened by climate change. This is not to give in to the view that we ought simply to aggregate the interests of the parties concerned. The view that the fundamental interests of individuals sometimes need to be traded-off against each other does not equal the view that it is permissible to trade-off the fundamental interest of the few against the trivial interests of the many (Waldron 1989). Since the present conflict is one between duties to respect fundamental interests, where trivial interests never enter the calculation, the exercise of a trade-off is perfectly legitimate within a rights-based framework.

What is the nature of the human rights interests involved? Duties of democracy correspond to the human right to democracy that is in turn grounded in instrumental considerations about the likely preconditions for the preservation of human rights. The evidence, as explained by Christiano, applies to interests in not being murdered, tortured or imprisoned for political reasons by the state or its supporters (Christiano 2009, 148). In the end, human rights to democracy are in other words taken to be necessary requirements for the protection of human rights to personal integrity.[17] The significance of these interests should now be compared with the interests grounding climate duties. The basis for such duties does not derive from the misuse of political power by public officials but by the risks posed by a degraded climate that is caused by a multiplicity of activities by all kinds of agents. More precisely, climate change is believed to undermine human rights to life, health and subsistence (Bell 2011; Caney 2010). The conflict between climate duties and democratic duties accordingly equals the claim that duties corresponding to human rights of personal integrity conflict with duties corresponding to human rights in life, subsistence and health.

The normative significance of the interests that justify these rights can be spelled out in different ways. According to one view, the fact of being subject to murder, torture or persecution does not just entail enduring physical and psychological pain but also compromises the capacity to respond to and act from moral considerations. In the words of James Griffin, torture constitutes a fundamental wrong not just because it is painful but because it constitutes 'an attack on normative agency' (Griffin 2008, 52). However, it appears equally clear that normative agency is imperiled when life, subsistence or health is compromised. Actions that undermine the physical preconditions for moral reflection are accordingly undermining the same fundamental interests that are violated by murder, torture and persecution. If human rights are ultimately justified by reference to our interest in normative agency, it follows that the same basic interests are at play whether we speak of human rights in the context of climate change or in the context of authoritarian governments. Given that climate duties and democratic duties serve the same fundamental interests, there is no basis for determining their relative importance with respect to the interests at stake. The image of a conflict between these duties is generated, not by the fact that they protect different kinds of interests, but by the fact that they protect human rights from different kinds of threats.

## Institutional Solutions

Let us briefly consider a different method for resolving the conflict. When two or more moral demands conflict we should pause to consider if the action required is in fact necessary in order to discharge our duties. In case some feasible alternative is available, the conflict could potentially be avoided. The logic of this argument derives from the proportionality test that is increasingly applied by courts in cases where the question concerns the legitimacy of infringements of individual rights. A fundamental precept of the proportionality test is that intrusions on rights can be legitimate only if the intrusion is indeed 'necessary' in order to protect legitimate interests (Barak 2012, 317ff.).

So the relevant question is whether the protection of human rights from the consequences of climate change necessarily requires infringements of the human right to democracy. In this regard, we might note an interesting asymmetry in the relationship between climate duties and democratic duties with respect to the type of threat they are intended to counter. In case of democratic duties, the rationale is that human rights are more vulnerable where political institutions deny citizens opportunities to participate in collective decisions as this would weaken the political incentives of rulers to respect citizens' human rights. In brief, we have duties to create institutions that contribute to the creation of incentives for *others* not to violate human rights. The rationale looks very different in the case of climate duties. The reason why we cannot completely rule out climate duties to support non-democratic institutions is that democratic institutions may not be effective in preventing dangerous climate change from materializing. But since public policy in democratic institutions depends on public opinion, the incapacity of democratic institutions to tackle climate change must at least partly be explained in terms of public opinion. Thus, climate duties call for the creation of institutions that contribute to the creation of incentives for *us* not to violate human rights.

The point is that climate duties do not *necessarily* require the suspension of democratic institutions in order to alleviate climate change since it is possible for citizens to act within democratic institutions to create support for climate-friendly policies. Although there might be strong political incentives not to act in this way, it is possible for us to resist these incentives. Hence, the introduction of other political institutions in order to generate policies consistent with the protection of human rights from the consequences of climate change is not necessary in order for the realization of such policies. The conclusion is thus that democratic duties, but not climate duties, are necessary preconditions for the protection of human rights.[18]

In addition, the claim that climate duties potentially justify support for institutions that are not democratic also depends on factual premises about the empirical relationship between democratic institutions and climate policy. Studies comparing democratic and non-democratic political regimes with respect to climate policy are still rare and the results produced are somewhat ambiguous. According to one highly cited study, there is ample evidence to support the contention that democracies show greater determination in committing to reduce greenhouse gas emissions, whereas there is little evidence that emissions are in fact reduced (Bättig and Bernauer 2009). On the other hand, several other studies lend support to the tenet that democracies are more successful in reducing greenhouse gas emissions and in responding to other environmental challenges (e.g., Lachapelle and Paterson 2013; Li and Reuveny 2006).

However, once we are concerned mainly with the preservation of the human rights of present citizens, we should pause to consider what is to be counted as relevant evidence. Indeed, failure to launch effective policies to mitigate climate change does not necessarily seem pertinent. In order to protect present citizens from harm, we should rather focus on the capacity of governments to implement effective policies of adaptation and disaster risk reduction. Given that we are concerned exclusively with the human rights of contemporaries, the question is not whether future dangerous climate change can be avoided but whether adequate protection from the effects of climate change in the near future can be counted on. Although evidence is incomplete, studies focusing on 'adaptation through vulnerability reduction' count access to political rights and the ability of citizens to hold governments accountable for failure to protect them as among the decisive factors (Brooks, Adger and Kelly 2005). Empirical evidence thus tentatively confirms that the human rights of citizens in democratic nations are more likely to be protected from dangerous climate change than are the human rights of citizens in non-democratic nations. In sum, the notion that tensions between climate duties and democratic duties exist can potentially be bridged by democratic institutions.

## INTERGENERATIONAL DUTIES

The previous analysis appears unduly limited, however, once we consider the nature of the threat to human rights that grounds climate duties. Climate change is a problem of gigantic proportions, extending both in space and time and is aptly characterized as both 'genuinely global' and 'strongly intergenerational' (Gardiner 2013, 212). This means that its potential victims are to be found not only among contemporaries but also among future people and perhaps even more often so. In exploring the relationship between democratic duties and climate duties, it would be disastrously misplaced to confine ourselves to contemporaries without considering duties to future generations. Hence, the potential conflict between democratic duties and climate duties needs to be addressed from an intergenerational perspective.

To begin, we should ask whether the conclusion reached in the previous section—that conflicting duties towards contemporaries can be accommodated by democratic institutions—is equally valid with regard to intergenerational duties. Can we confidently submit that conflicts between climate duties and democratic duties towards future people can be avoided by rendering support to democratic institutions?

Just as in the intragenerational case, two considerations determine the extent of conflict between climate duties to future people and democratic duties to future people. First, there must be reason to believe that democratic institutions are instrumental to the protection of future people's human rights from the consequences of climate change. If this can be ascertained, climate duties towards future people would seem to require support for democratic institutions, thus avoiding any conflict with democratic duties to future people. Second, we need grounds to believe that future people are entitled to democratic government as a matter of human rights. The validity of this claim is necessary, albeit not sufficient, in order for there being a conflict between the human rights duties towards future people. In case future people are not entitled to democratic institutions as a matter of human rights, our democratic duties towards future people appear overridden by climate duties because the latter are required for the preservation of human rights and human rights do have priority.

Conflicting human rights duties towards future people consequently exist only if the following claims are both true: (i) democratic institutions are not instrumental to the protection of future people's human rights from the consequences of climate change and (ii) we have democratic duties towards future people grounded in the preservation of their human rights. The conjunction of these claims represents a dilemma since it entails human rights duties towards future people both to support democratic institutions and to support alternative political institutions with the capacity to reduce the impact of climate change on future generations. We now proceed to evaluate these claims in turn.

## PROTECTING FUTURE INTERESTS

Consider first the claim that democratic institutions are instrumental to the preservation of future people's human rights in face of dangerous climate change. In order for the question to be meaningful we must assume that current atmospheric concentrations of greenhouse gases are not already too high to prevent dangerous climate change in the future to the extent associated with significant violations of future people's human rights. The UN Framework Convention on Climate Change was premised on the notion that future increases in global mean temperatures must not exceed two degrees Celsius in order to safely avoid dangerous climate change from unfolding. Yet, given the paucity of climate policy since these targets were first identified, avoiding these temperatures is no longer feasible. Four degrees' increase in global mean temperatures should now be expected on current trajectories (Rogelj et al. 2009; New et al. 2011). On the assumption that increasing global mean temperatures of four degrees could potentially be avoided, given radical shifts in public policy, and that future people's human rights will not be seriously compromised below this level, the question is whether democratic governments can be expected to adopt the necessary policy changes.

Positive evidence to answer this question with confidence is barely available. Yet, multiple voices argue that we have little reason to ever expect democracy to adequately address long-term risks. The reason is that democratic institutions offer strong incentives for governments to attend to the short-term interests of the living citizens whereas incentives to accommodate the interests of future people are weak. Because of the short-termism encouraged by electoral cycles and the fact that only the living generations are enfranchised, democracy 'systematically undervalues the interests of future generations' (Boston and Lempp 2010, 1004; Tremmel 2013).[19] Democracy is accordingly depicted as biased towards the present, triggering what is essentially a 'tyranny of the contemporary' (Thompson 2010; Gardiner 2014). Of course, it is not necessary to subscribe to the claim that democracy is *generally* biased against long-term interests in order to conclude that democracy is biased against the interests of future people in protections against climate change. For the latter claim to be true, it is sufficient to accept that democracy is biased with respect to *specific* long-term interests.

Moreover, it is undoubtedly controversial to argue that features inherent to democracy practically ensure that the interests of future generations will be ignored. The real world of democracy should not be confused with the democratic ideal (Dahl 2007). Hence, the presentist bias of democracy as it is currently practiced does not entail that the democratic ideal is biased against the future. The expectation that 'democratizing' the institutions of electoral democracy would eliminate presentist biases is reflected in claims to 'greening' liberal democracy. By strengthening the deliberative element in democratic procedures effective climate policy should be more likely to garner political support (e.g., Barry, 2001; Broome 2008; Bäckstrand et al. 2010). The proposed changes may be radical indeed. Following Dryzek and others, deliberative democracy can only be expected to be successful once it is implemented at the global level. Hence, it is proposed that current political institutions at the level of the nation-state are replaced by global democracy modelled on the idea of 'earth system governance' (Stevenson and Dryzek 2014).

However, the view that we ought to revise rather than replace democracy in order to avoid its 'presentist' biases has troubling implications with respect to the human rights of contemporaries. Whereas responsiveness to the interests of contemporaries appears worrying from an intergenerational perspective, it seems likely that the positive empirical relationship between democratic institutions and the human rights of contemporaries depends exactly on the responsiveness of such institutions to the needs and complaints of their citizens. As previously argued by Amartya Sen, Henry Shue and others, rights to political participation create powerful 'political incentives' of rulers to attend to citizens' fundamental interests (Sen, 1999; Shue, 1996). The consequence of weaker presentist biases might consequently be that protections for the human rights of contemporaries are weakened as it also weakens the political incentives of rulers to attend to their interests. In the end, the attack on presentism could inadvertently undermine the rationale for

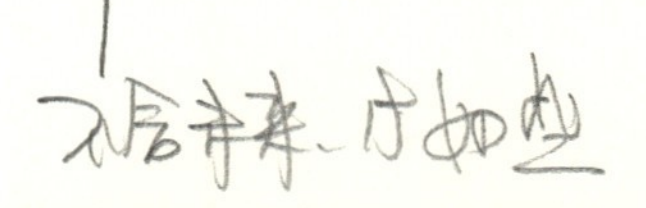

the human right to democracy as it depends on the instrumental efficacy of democratic institutions to protect the human rights of contemporaries.[20]

## FUTURE RIGHTS TO DEMOCRACY?

Now, let us consider the claim that the human rights of future people include democratic rights. According to this view, the human rights of future people are not only grounded in their interests in life, subsistence and health as they derive also from their interests in democratic forms of government. Our democratic duties towards future generations would in other words be human rights duties.

An argument to this effect is that the considerations relevant in deciding the human rights of the living must also be relevant in deciding the human rights of the yet unborn. Hence, in case we accept that existing people do have human rights to democracy, we ought to accept that future people have human rights to democracy as well. This view is not premised on the notion that human rights are diachronic in the sense that the human rights recognized as valid for human beings today must also be recognized as valid for human beings alive at any time (Bell 2011, 107). The argument is rather that the considerations that contribute to justify human rights today would also contribute to justify the human rights of people in different times *to the extent that these considerations apply*.[21] Hence, future people hold human rights to democracy if considerations that justify contemporaries' human rights to democracy apply also to future people. The question is, of course, whether there is reason to believe that the exact considerations do apply in the future. If they do, human beings alive today would seem to have reason to take steps in order to protect the human right to democracy of future people.

Now, whenever we speak about the human rights of future people there is the immediate objection that people who do not exist could not be bearers of rights at all (De-Shalit, 1991, 108). Just as the person bearing the right is non-existent, the rights and other attributes of this non-existing person are non-existent. On the other hand, if there will be future people, there is no objection against us thinking that they will have rights. Thus, as pointed out by Axel Gosseries and others, the non-existence objection does not preclude recognition of the *future* rights of future people (Gosseries 2008; Elliot, 1989). The question is rather whether the future rights of future people create corresponding duties among the living.

The answer to this question depends on the likelihood that future people with future rights will exist. In case it is just remotely probable that future people will ever come into existence, the future rights of future people will be equally improbable. It nevertheless appears plausible to assume that the coming into existence of future people represents a 'significant probability'

(Gosseries 2008, 456; Bell 2011, 108). The rationale for concluding that we currently have duties to secure the conditions for future people's future rights is consequently that we have reason to be confident in the future existence of future people.

This is all very well. The problem then is just that the future human right to democracy of future people depends on additional conditions. The justification of the human right to democracy depends on the validity of empirical generalizations about the effects of democratic institutions. Estimates of the causal effect of democratic institutions on human rights are generated on the basis of observations of and controls for a wide range of political and economic facts that obtain in present circumstances. The point is that all these features of the world together define the preconditions for the positive 'net effect' of democracy on the protection of human rights (e.g. Mahoney 2008, 422).

Yet, predictions of institutional outcomes in the future are considerably less certain than predictions of the coming into existence of future people. The causal relationship between democratic institutions and human rights in the distant future is conditioned by the same variety of political and economic facts that condition the estimated causal effect today. The difference is, of course, that these political and economic facts will be very different. The *future* 'net effect' of democratic institutions on human rights is therefore much less certain than in present circumstances. This is important because if duties corresponding to the future rights of future people depend on the 'significant probability' of the coming into existence of the *conditions* for future rights, it follows that there is little evidence to support future human rights to democracy of future people. This point does not exclude the coming into existence of evidence that grounds the future human rights to democracy of future people. But it suggests that such grounds are harder to identify than grounds that justify human rights to democracy of contemporaries. What follows from these considerations is that instrumental justifications for human rights to democracy are less credible with respect to future people than with respect to contemporaries. In fact, these considerations also apply to claims concerning the ability of democratic institutions to protect the human rights of citizens from the effects of climate change. Despite tentative evidence to conclude that democracies are able to muster effective policies of adaptation and disaster risk reduction, the same evidence does not authorize confidence that the complex relationship valid under present conditions is valid also in the distant future.

The overall conclusion is that human rights are unlikely to offer a credible foundation for democratic duties to future generations. This does not deny other grounds for such duties. It may, for example, be held that each generation has the right to exercise popular sovereignty and that every generation therefore has a claim against previous generations to the constitutional preconditions for the exercise of this right (e.g., Thompson 2010). The point is just that democratic duties towards future people are not grounded in future *human rights*.

The implication for our duties towards future generations should be clear. On the basis of the tenet that human rights have priority, climate duties towards future human rights should be granted priority before democratic duties towards future generations. In order to discharge our duties towards future generations we should in other words be prepared to lend support to the creation of political institutions with the capacity to protect future human rights even if such institutions do not satisfy the criteria of procedural democracy.

On the other hand, the nature of our duties *overall* must depend on an integrated account that takes into consideration duties towards both future generations and contemporaries. Following the argument of the previous section, duties towards contemporaries include climate duties as well as democratic duties, where both sets of duties are understood as grounded in human rights. Given that democratic institutions are instrumental to the preservation of human rights among contemporaries it consequently follows—as already argued—that our duties towards contemporaries are best secured by rendering support to democratic institutions. The troubling result is that our overall duties with regard to human rights entail duties to support democratic institutions in order to secure the human rights of contemporaries whereas there is no basis for corresponding democratic duties with respect to future generations.[22]

## CONCLUSIONS

In this chapter we have explored some implications of the tenet that individuals have climate duties that are triggered by the expected consequences of climate change and that such duties are particularly weighty as they are generated by the imperative to secure human rights. The view that such duties exist invites potentially radical conclusions as it entails that we have duties also to support the institutions necessary for the protection of human rights. Yet, since it is often believed that we have duties towards democratic institutions too, the relevant question to ask is how climate duties relate to duties towards democratic institutions. The question appears particularly challenging once we accept that human rights have priority and that duties to respect human rights from the effects of climate change therefore override other duties citizens may have.

The answer is shown to be contingent on whether we are addressing the human rights of contemporaries or the human rights of future people. This turns out to be crucial because there is reason to believe that democratic duties should be understood as human rights duties in relation to contemporaries but not in relation to future people. Hence, the priority of human rights is decisive in identifying climate duties to future generations but plays no significant role in relation to contemporaries. With respect to contemporaries, the evidence suggests that democratic duties and climate duties point in similar directions. The human rights of contemporaries are best served by democratic institutions, whether the threat is posed by climate change or by authoritarianism.

This result indicates that a commitment to climate duties requires attention to the democratic qualities of political institutions. The institutional alternatives on the menu are subject to the constraint that they must be consistent with the protection of the human rights of both present and future people. Exactly how these constraints should be understood remains to be specified, however. As far as we know, the requirement to protect human rights from the consequences of climate change is presenting us with a potential conflict between the democratic duties owed to contemporaries and the climate duties owed to future generations, and it is not clear that we will be able to discharge those duties consistently.

## REFERENCES

Aminazadeh, Sara C. 2006. 'A Moral Imperative: The Human Rights Implications of Climate Change'. *Hastings International and Comparative Law Review* 30: 231–65.

Ashford, Elisabeth. 2006. 'The Inadequacy of Our Traditional Conception of the Duties Imposed by Human Rights'. *Canadian Journal of Law and Jurisprudence* 19(2): 217–35.

Bäckstrand, Karin, et al. Editors. 2012. *Environmental Politics and Deliberative Democracy*. Cheltenham: Edwar Elgar.

Baer, Paul, and Ambuj Sagar. 2009. 'Ethics, Rights and Responsibilities'. In *Climate Change Science and Policy*, edited by Stephen Schneider et al. Washington, DC: Island Press.

Barak, Ahron. 2012. *Proportionality. Constitutional rights and their limitations*. Cambridge: Cambridge University Press.

Barry, John. 2001. 'Greening Liberal Democracy: Theory, Practice and Political Economy'. In *Sustaining Liberal Democracy: Ecological Challenges and Opportunities*, edited by John Barry and Marcel Wissenburg, 59–81. London: Palgrave.

Beckman, Ludvig. 2014. 'The Right to Democracy and the Human Right to Vote: The Instrumental Argument Rejected'. *Journal of Human Rights* 13: 381–94.

Beitz, Charles. 2001. 'Human Rights as a Common Concern'. *The American Political Science Review* 95: 269–82.

Bell, Derek. 2011. 'Does anthropogenic climate change violate human rights?' *Critical Review of International Social and Political Philosophy* 14: 99–124.

Bodansky, Daniel. 2010. 'Introduction: Climate Change and Human Rights: Unpacking the Issues'. *Georgia Journal of International and Comparative Law* 38: 511–24.

Boston, Jonathan and Lempp, Frieder. 2011. 'Climate Change. Explaining and solving the mismatch between scientific urgency and political inertia'. *Accounting, Auditing & Accountability Journal* 24: 1000–21.

Brandstedt, Eric and Anna-Karin Bergman. 2013. 'Climate Rights—Feasible or Not?'. *Environmental Politics* 22: 394–409.

Brink, David O. 1994. 'Moral Conflict and Its Structure'. *The Philosophical Review* 103: 215–47.

Brooks, Nick; W. Neil Adger, P. Mick Kelly. 2005. 'The determinants of vulnerability and adaptive capacity at the national level and the implications for adaptation'. *Global Environmental Change* 15: 151–63.

Broome, John. 2008. 'The Ethics of Climate Change'. *Scientific American* 298: 97–102.

Buchanan, Allen. 2005. 'Equality and Human Rights'. *Politics, Philosophy and Economics* 4: 69–90.

Caney, Simon. 2010. 'Climate change, human rights and moral thresholds'. In *Human Rights and Climate Change*, edited by Stephen Humphrey and Mary Robinson, Cambridge: Cambridge University Press.

Christiano, Thomas. 2009. *The Constitution of Equality*. Oxford: Oxford University Press.

———. 2011. 'An instrumental argument for a Human Right to Democracy'. *Philosophy and Public Affairs* 39: 142–76.

Cripps, Elisabeth. 2013. *Climate Change and the Moral Agent*. Oxford: Oxford University Press.

Cruft, Rowan. 2012. 'Human Rights as Rights'. In *The Philosophy of Human Rights*, edited by Gerhard Ernst and Jean-Christophe Heilinger. Berlin: De Gruyter.

Dahl, Robert. 2007. *On Political Equality*. Cambridge, MA: Harvard University Press.

De-Shalit, Avner. 1991. 'Community and the Rights of Future Generations: A Reply to Robert Elliot'. *Journal of Applied Philosophy* 9: 105–15.

Elliot, Robert. 1989. 'The Rights of Future People'. *Journal of Applied Philosophy* 6: 159–70.

Franck, Thomas M. 1992. 'The Emerging Right to Democratic Governance'. *The American Journal of International Law* 86: 46–91.

Freeman, Michael. 2004. 'The Problem of Secularism in Human Rights Theory'. *Human Rights Quarterly* 26: 375–400.

Gardiner, Stephen. 2006. 'A Perfect Moral Storm: Climate Change, Intergenerational Ethics and the Problem of Moral Corruption'. *Environmental Values* 15: 397–413.

———. 2013. 'Human Rights in a Hostile Climate'. In *Human Rights: The Hard Questions*, edited by Cindy Holder and David Reidy. Cambridge: Cambridge University Press.

Gilalbert, Pablo. 2011. 'Humanist and Political Perspectives on Human Rights'. *Political Theory* 39: 439–67.

Gosseries, Axel. 2008. 'On Future Generations' Future Rights'. *The Journal of Political Philosophy* 16: 446–74.

Green, Maria. 2001. 'What We Talk About When We Talk About Indicators: Current Approaches to Human Rights Measurement'. *Human Rights Quarterly* 23: 1062–97.

Griffin, James. 2008. *On Human Rights*. Oxford: Oxford University Press.

Hayward, Tim. 2012. 'Climate Change and Ethics'. *Nature Climate Change* 2: 843–48.

Howarth. Richard. 2011. 'Intergenerational Justice'. In *The Oxford Handbook of Climate Change and Society*, edited by John S. Dryzek, Richard B. Norgaard and David Schlosberg. Oxford: Oxford University Press.

IPCC. 2014. *Climate Change 2014: Impacts, Adaptation and Vulnerability. Volume I: Global and Sectorial Aspects*.

Johnson, Baylor L. 2003. 'Ethical Obligations in a Tragedy of the Commons'. *Environmental Values* 12: 271–87.

Joseph, Sarah. 2010. 'Civil and Political Rights'. In *International Human Rights Law: Six Decades after the UDHR and Beyond*, edited by Mashood A. Baderin and Manisuli Ssenyonjo, 317–36. Farnham, UK: Ashgate Publishers.

Kamm, F. M. 2001. 'Conflicts of Rights: Typology, Methodology, and Nonconsequentialism', *Legal Theory* 7: 239–55.

Kates, Michael. 2013. 'Justice, Democracy and Future Generations'. *Critical Review of International Social and Political Philosophy*, DOI: 10.1080/1398230.2013.861655

Knox, J. H. 2009. 'Climate Change and Human Rights Law'. *Virginia Journal of International Law* 50: 163–218.

Lachapelle, Erick, and Matthew Paterson. 2013. 'Drivers of National Climate Policy'. *Climate Policy* 13: 547–71.

Lardy, Heather. 2003. 'Translating Human Rights into Moral Demands on Government'. *International legal theory* 9: 123–34.

Lemons, J., Brown, D.A. 2011. 'Global Climate Change and Non-violent Civil Disobedience'. *Ethics in Science and Environmental Politics* 11: 3–12

Li, Quan, and Rafael Reuveny. 2006. 'Democracy and Environmental Degradation'. *International Studies Quarterly* 50: 935–56.

Mahoney, James. 2008. 'Toward a Unified Theory of Causality'. *Comparative Political Studies* 41: 412–36.

McInerney-Lankford, Siobhan, Mac Darrow, and Lavanya Rajamani. 2009. *Human Rights and Climate Change. A Review of the International Legal Dimensions,* Washington, DC: The World Bank.

Mesquita, Bruce, Chreif Bueno, Marie Feryal, George Downs, and Alastair Smith. 2005. 'Thinking Inside the Box: A Closer Look at Democracy and Human Rights'. *International Studies Quarterly* 49: 439–458.

Mieth, Corinna. 2012. 'On Human Rights and the Strength of Corresponding Duties'. In *The Philosophy of Human Rights*, edited by Gerhard Ernst and Jan-Christophe Heilinger, Berlin: De Gruyter.

Munck, Gerardo L. 2009. *Measuring Democracy: A Bridge Between Scholarship and Politics*. Baltimore: Johns Hopkins University Press.

New, Mark, Diana Liverman, Heike Schroder, Kevin Anderson. 2011. 'Four Degrees and Beyond: The Potential for a Global Temperature Increase of Four Degrees and Its Implications'. *Philosophical Transactions of the Royal Society of London A: Mathematical, Physical and Engineering Sciences,* 6–19; DOI: 10.1098/rsta.2010.0303.

Nickel, James. 1993. 'How Human Rights Generate Duties to Protect and Provide'. *Human Rights Quarterly* 15: 77–86.

———. 2010. *Human Rights*. Stanford Encyclopedia of Philosophy: http://plato.stanford.edu/

Nihlén Fahlquist, Jessica. 2009. 'Moral Responsibility for Environmental Problems—Individual or Institutional?' *Journal of Agricultural Environmental Ethics* 22: 109–24.

Pogge, Thomas. 2009. 'Shue on Rights and Duties'. In *Global Basic Rights*, edited by Charles Beitz and Robert Goodin, 113–30, Oxford: Oxford University Press.

Raterman, Ty. 2012. 'Bearing the Weight of the World: On the Extent of an Individual's Environmental Responsibility'. *Environmental Values* 21: 417-436.

Rawls, John. 1971. *A Theory of Justice*. Oxford: Oxford University Press.

Raz, Joseph. 2010. 'Human Rights in the Emerging World Order'. *Transnational Legal Theory* 1: 31–47.

Reidy, David. 2012. 'On the Human Right to Democracy: Searching for Sense without Stilts'. *Journal of Social Philosophy* 43: 177–203.

Rogelj, Joeri, Bill Hare, Julia Nabel, Kirsten Macey, Michel Schaeffer, Kathleen Markmann, and Malte Meinshausen. 2009. 'Halfway to Copenhagen, No Way to 2 °C'. *Nature Reports Climate Change* , Doi:10.1038/climate.2009.57

Sen, Amartya. 1982. 'Rights and Agency'. *Philosophy & Public Affairs* 11: 3–39.

———. 1999. *Development as Freedom*. Oxford: Oxford University Press.

Shearman, David J. C., and Joseph Wayne Smith. 2007. *The Climate Change Challenge and the Failure of Democracy*. Westport, CT: Praeger.

Shue, Henry. 1996. Basic Rights. Subsistence, Affluence and US Foreign Policy, 2nd edition. Princeton: Princeton University Press.

Sinnott-Armstrong, Walter. 1996. 'Moral Dilemmas and Rights'. In Moral Dilemmas and Moral Theory, edited by Homer E. Mason. New York: Oxford University Press.

Sinnott-Armstrong, Walter. 2005. 'It's Not My Fault'. In *Perspectives on Climate Change*, edited by Walter Sinnott-Armstrong and Richard Howarth. Amsterdam: Elsevier.

Steiner, Henry. 1988. 'Political Participation as a Human Right'. *Harvard Human Rights Yearbook* 1: 77–134.

———. 2008. 'Two Sides of the Same Coin? Democracy and International Human Rights'. *Israeli Law Review* 41: 445–76.

Stevenson, Hayley, and John Dryzek. 2014. *Democratizing Global Climate Governance*. Cambridge: Cambridge University Press.

Tasioulas, John. 2010. 'Taking Rights out of Human Rights'. *Ethics* 120: 647–78.

Thompson, Dennis. 2010. 'Representing Future Generations: Political Presentism and Democratic Trusteeship'. *Critical Review of International Social and Political Philosophy* 13: 17–37.
Tremmel, Joerg Chet. 2013. 'Climate Change and Political Philosophy: Who Owes What to Whom?' *Environmental Values* 22: 725–49.
Tunick, Mark. 1998. 'The Scope of Our Natural Duties'. *Journal of Social Philosophy* 29: 87–96.
Valentini, Laura. 2012. Ideal vs. Non-ideal Theory: A Conceptual Map'. *Philosophy Compass* 7: 654–64.
Wellman, Carl. 1995. 'On Conflicts Between Rights'. *Law and Philosophy* 14: 271–95.
Wolfe, Matthew. 2008. 'The Shadow of Future Generations'. *Duke Law Journal* 57: 1897–1932.

## NOTES

1. I am very grateful for instructive comments on earlier versions of this paper from the editors of this volume and from the participants in the political theory session of the Nordic Political Science Association's meeting in Gothenburg, August 2015.

2. The claim that 'climate rights' are implied by human rights has been questioned by reference to the fact that no legitimate 'formal structure' exists that recognizes such rights (Brandsted & Bergman, 2013, 400). However, in this chapter human rights refer to morally required human rights. The existence of human rights so understood is not conditioned by the existence of 'formal structures' that recognize their validity.

3. Shue refers not to human rights but to 'basic rights'. However, the distinction is primarily to remind us that legally recognized human rights do not equal the 'genuine' or 'proper' set of morally justified human rights (see Beitz, 2001, 272). In what follows I use the terminology of human rights as I find this customary even when we speak about morally justified basic rights.

4. By contrast, international climate agreements exclusively refer to the responsibilities of 'nations'. See Knox (2009) for a discussion.

5. In addition, climate duties, as specified by Bell, may require international political institutions. In this spirit, Elisabeth Cripps argues that dangerous climate change generates 'promotional duties' that include the duty to support institutions at the international level 'if necessary to do the job' (Cripps, 2013, 147).

6. I will speak interchangeably about conflicts between rights and duties. When rights conflict they do so only because they generate conflicting duties (Waldron, 1989, 511).

7. Some would deny this possibility. They believe justified rights-claims cannot be overridden unless they are in fact unjustified rights-claims (e.g., Wellman, 1995, 275).

8. See Tunick (1998, 88) for the distinction between duties to 'support' and to 'comply with' democratic decisions.

9. For other senses of the distinction between ideal and non-ideal conditions, see Valentini (2012).

10. For a critique of this view, see Freeman (2004).

11. Not all human rights are grounded in the fundamental importance of individual interests. However, for present purposes I ignore human rights derived from the collective interests (e.g., the human right to self-determination).

12. The distinction between 'positive' and 'negative' duties is not necessarily equal to the distinction between duties that require positive action and duties that require not acting in certain ways. An alternative is to conceive of negative duties as requirements to respect human rights and positive duties as requirements to *improve* respect for human rights. Following the latter terminology, negative duties undoubtedly require positive action whenever necessary for respecting human rights, whereas positive duties sometimes require not acting at all whenever necessary for the improvement of respect for human rights. See Mieth (2012, 167).

13. Legal human rights to democracy exist if recognized in the international legal system (i.e., in treatises, case-law or custom). Although human rights to political participation and

voting are recognized therein, they were not intended at the time of their adoption, nor have they subsequently been understood, as tantamount to a general right to democratic government (Steiner 1988; Lardy 2003; Joseph 2010). A different possibility is to point at prevalent commitments of international organizations to promote democracy and 'rights to democratic governance'. Yet, again, these statements are more appropriately understood as reporting the aspirations of the organizations rather than as legally binding entitlements. In sum, while democracy is not yet fully recognized as part of human rights law, democracy does appear to be an 'emerging right' that lends support from 'tendencies' in international law (Franck 1992, 71; Steiner 2008).

14. The general structure of the instrumental argument for human rights to political participation is delineated by Shue (1996, 74ff). An alternative approach is to insist that democratic rights protect fundamentally important interests of the rights-bearer and should therefore be considered human rights for intrinsic reasons (e.g., Cruft, 2012, 133). The problem facing this approach is to explain why political participation is among the prerequisites for 'minimally decent life'. Unless this can be made credible, democracy may as well be conceived of as 'universal justice-based moral rights' instead of as 'human rights' (Reidy 2012, 201).

15. A good overview of empirical evidence is found in Mesquita, Chreif, Downs and Smith (2005).

16. For doubts about the conclusiveness of instrumental justifications of human rights to democracy, see Beckman (2014).

17. It might seem relevant here to invoke Amartya Sen's (1999, 152) famous argument that democratic government is superior in terms of protecting citizens from starvation and other deprivations. However, Sen does not explicitly suggest that this consideration justifies human rights to democracy. The empirical relationship pin-pointed by Sen might nevertheless be pertinent for the assessment of the instrumental value of democracy given the aim to combat climate change—as will appear from the following discussion.

18. As noted by Shue (1996, 87) the 'necessity' of democracy for the protection of human rights depends on empirical generalizations based on historical evidence and therefore should not be understood as a theoretical proposition ruling out as 'utterly inconceivable' that institutions without participation would secure human rights.

19. 'Presentism' is also used as a description of the framework implicit to climate economics according to which policies should be based only on the preferences of the living generations (Howarth, 2011).

20. A further objection is the point argued by Kates (2013) that the bootstrapping problem is inherent to claims to the effect that democratic institutions should be reformed in order to protect present and future interests from climate change. That is, such arguments depend on the conflicting assumptions that (a) democratic institutions constitute barriers to effective climate policy and (b) people can act within such institutions to remove those barriers. Of course, the truth of (b) does appear to speak against the truth of (a) and vice versa.

21. Raz (2010) argues that human rights are only synchronic (i.e.. valid for human beings alive today). It would be 'absurd', Raz argues, to claim that Neolithic human beings had a human right to education just because we today recognize such a human right. My point is that considerations that justify human rights, including the human right to education, are valid diachronically but that the applicability of these considerations at different times cannot be taken for granted.

22. It might be objected that we have reason to support democratic institutions for future people since we have already accepted that they more effectively provide for effective policies of adaptation and disaster risk reduction. But if uncertainty about future causal relationships undermines the instrumental argument for future human rights to democracy, it should undermine also the argument that democracy is instrumental to the protection of other future human rights. Despite tentative evidence that democracies are able to muster effective policies of adaptation and disaster risk reduction, the evidence does not authorize confidence that the complex relationship valid under present conditions is valid also in the distant future.

*Chapter Four*

# Gridlock in Global Climate Change Negotiations[1]

## *Two Democratic Arguments against Minilateralism*

Jonathan W. Kuyper

### INTRODUCTION

In 1992, national governments across the globe negotiated the United Nations Framework Convention on Climate Change (UNFCCC) to tackle the growing issues of anthropogenic climate change. Since the introduction of the Kyoto Protocol in 1997, various attempts have been launched to create another legally binding treaty for regulating global warming. At the thirteenth Conference of the Parties (COP) in Bali during 2007 it was decided that all major emitters would sign a legal treaty two years later at COP 15 in Copenhagen. Again at COP 17 in Durban during 2011, COP 21 in Paris was set as the timeline for a new multilateral agreement. Yet despite these attempts, no binding agreement has come to pass and the prospects for Paris in 2015 remain uncertain at best. As such, many commentators have suggested that the UNFCCC system is incapable of producing the kind of binding treaty required to meet climate change problems (Naím 2009; Grasso & Roberts 2014; Jamieson 2014).

The failure to produce a binding outcome is often referred to as negotiation gridlock. Generally speaking, two responses have been formulated: inclusive multilateralism and exclusive minilateralism (Eckersley 2012, 25). Inclusive multilateralism stresses the benefits of procedural legitimacy: all sovereign states should take part in a communicative and dialectic process of negotiation to help shape global rules. Although proponents recognize that gridlock is problematic, they maintain that inclusive and democratic procedures are vital to maintain decision-making legitimacy. In contrast, exclusive minilateralism rests on a logic of effectiveness and feasibility. Proponents contend that a consensus decision among 194 states on an issue as politicized and complex as climate change is a chimera (Victor 2011). Given the need for action on climate change as soon as possible, advocates suggest that the best way to advance negotiations is for a small club of high emitters to take the lead and negotiate a treaty to reduce emissions (Keohane and Victor 2011, 10). Intermediary positions, such as 'inclusive minilateralism' by Robyn Eckersley (2012) have attempted to reconcile proceduralism with pragmatism by limiting climate change negotiations to the states which are most capable, responsible and vulnerable.

Minilateral arguments for countering negotiation gridlock thus center on the claim that a small club negotiation will help advance international action to mitigate climate change in both an *effective* and *timely* manner. In this chapter I argue against minilateralism. I contend that climate change negotiations should remain within the UNFCCC system and strive to maintain inclusivity of both state and non-state actors. This claim turns the argument of minilateralists against themselves by suggesting that exclusivity undermines the epistemic benefits and compliance required for an *effective* climate change policy. In other words, democratic inclusion in negotiations should be viewed as essential for producing meaningful outcomes. Given the minilateral emphasis on *effectiveness*, this is a strong counter argument. Because we still require a way to overcome gridlock in a *timely* manner, I also suggest that flexibility mechanisms—such as sunset provisions, relational contracts and especially escape clauses—should be employed to advance negotiations.

The chapter proceeds in four sections. First I outline in more detail the differences between inclusive multilateralism and exclusive minilateralism and make connections with input and output variants of legitimacy. Second, I suggest that multilateral negotiations should be democratic and tie this to instrumental considerations. I outline how epistemic benefits and compliance can be derived from inclusive multilateralism and how minilateral restrictions would undercut these outcomes. Third, I suggest some pragmatic ways that broad-scale multilateralism can be made more tractable by employing flexibility mechanisms and I discuss a range of examples which could help undergird this direction for research. The final section concludes.

## CLIMATE CHANGE NEGOTIATIONS: GRIDLOCK AND DEMOCRACY

### Inclusive Multilateralism vs. Exclusive Minilateralism

The network of actors and institutions that seek to govern climate change at the global level is dense and fragmented (Zelli 2011). This system is composed of a patchwork of state and transnational non-state actors (TNAs) that have forged legal regimes, expert panels, adaption initiatives, bilateral deals, public-private partnerships and private organizations (Abbott 2012; Bäckstrand 2008). At the center of efforts to manage and mitigate climate change is the multilateral UNFCCC process. The UNFCCC, as with all multilateral negotiations, rests upon recognized principles of sovereignty and reciprocity between states (Ruggie 1992). By and large, these principles are supposed to ensure that multilateral negotiations give all states the ability to voice their opinion, jointly author international laws and ratify agreements of which they approve.[2]

Although the UNFCCC has produced some tangible outputs—most notably the Kyoto Protocol—recent years have witnessed a rise in protests, politicization and pushback as negotiations have stalled and reform proposals have been side-stepped. At COP 15 in Copenhagen, negotiations failed to produce anything resembling a treaty to limit global warming. The reasons for this are certainly numerous, ranging from China and Saudi Arabia's intransigence, the US's small targets, the EU's waning leadership, the Danish host's poor diplomatic efforts and so on (Stevenson 2014).[3] The major output of COP 15 was a weak statement to 'take note of the Copenhagen Accord', produced when a small group of powerful states attempted to take matters into their own hands. In Warsaw four years later (COP 19), 132 states walked out of the climate negotiations because developed states sidelined and postponed issues of technology transfer and compensation. Repeated failures at COP meetings have gone hand-in-glove with a general intensification of international climate negotiations (Depledge and Chasek 2012, 19). Over the past 20 years, the number of negotiating bodies, meetings, participants and ministers in the decision-making process has risen sharply, yet substantive action on greenhouse gas emission and global warming has remained unobtainable.

This situation, as noted, has been described as negotiation gridlock. Many commentators have responded to this gridlock by stressing either the importance of multilateral inclusion or minilateral exclusion. The basis for these positions corresponds respectively with a distinction raised by Fritz Scharpf (1999) between input and output variants of legitimacy. Input legitimacy relies upon decision-making being open, equitable and fair for all participants. When decision-making processes uphold these criteria, legitimacy—specifically democratic legitimacy—is enhanced. Output legitimacy, though, is augmented through effective and efficient results being generated. By this token participants are more likely to find decision-making democratically legitimate if effective outcomes are reached in a timely manner irrespective of the qualities embedded in the decision-making process.

Inclusive multilateralists argue that input legitimacy is essential and hence tend to underscore the importance of procedural fairness and cosmopolitan inclusion. The basic idea underpinning this response rests upon something akin to the Habermasian notion of (deliberative) democracy in which all affected individuals (or their representatives) should be included in decision-making processes (Page 2011, 53; Eckersley 2012). Inclusion helps recognize and respect human autonomy as a fundamental normative good that all individuals are owed *qua* individuals. Because it is clearly not possible to involve seven billion individuals in the UNFCCC negotiations, sovereign states are supposed to represent their citizenry in negotiations. Individual autonomy is respected via the norm of sovereignty that provides states with a baseline of independence. Decision-making should be grounded in the use of public reason so that affected parties can recognize and accept the process and outcome.

The ideal of deliberative democracy entails several important principles which should be procedurally upheld: inclusivity, non-coercion and reciprocity. First, multilateral climate change debates should be open to as many competing discourses, ideas and sets of interests as possible. Moreover, negotiations should, wherever possible, be non-coercive so that 'the unforced force of the better argument' guides public policy and rule-making (Habermas 1993). Finally, deliberators should go beyond rational self-interest and seek to frame their claims in terms acceptable to others. These procedural ideals help to foster democratic and legitimate negotiations which respect the autonomy of individuals.

Critics of the gridlocked UNFCCC process have often suggested that the time-sensitive nature of climate change means that input legitimacy should be sacrificed in favour of reaching effectual outcomes sooner rather than later. Inclusive deliberation, transparency, consensus-building and accountability—all elements of input legitimacy and procedural fairness—hobble the negotiation process and make reaching agreement impossible. Indeed, the lack of agreement on an issue such as climate change is also argued to be morally problematic as individuals suffer the consequences of rising temperatures, sea levels and $CO_2$ emissions. Instead of upholding inclusive multilateralism, then, the number of negotiating states should be limited to make reaching agreement tractable. This could enhance the output dimension of global climate change negotiations by helping reach an effective agreement faster.

David Victor (2009; 2011), in a series of books and articles, has been an ardent supporter of exclusive minilateralism to break gridlock. Although Victor—along with Moisés Names, Anthony Giddens and others—concedes that a minilateral club model is both exclusionary and procedurally undemocratic, they emphasize that a small group of key states are best equipped to reach an effective agreement and keep global warming below 2°C of pre-industrial levels. Names suggests that around 20 states constitute the 'magic number' for an effective and impactful agreement. Victor (2011, 276) contends that a small group of major emitters are best placed to avoid the burden of consensus multilateralism. Turning toward international trade negotiations for precedent, Victor (2009, 343) argues that a smaller bloc of states can credibly commit to climate policy more effectively than can larger groups. Victor (2009, 344) suggests that the 17-member Major Economies Forum on Energy and Climate (MEF) or the International Energy Agency (IEA) comprised of 28 member governments represent useful models for replication.

In a slightly different vein, Eckersley has attempted to tread a middle path between exclusive minilateralism and inclusive multilateralism. Following the UNFCCC principle of common but differentiated responsibilities, Eckersley (2012, 36) advocates 'inclusive minilateralism' which seeks to reduce the number of negotiating parties by only including the most 'capable', 'responsible' and 'vulnerable' states. The most capable states are those with the highest GDP, the most responsible states are the historically biggest emitters and the most vulnerable states refer to representatives from small island governments, the African Group (AG) and the least developed countries (LDC). This would reduce the number of negotiating partners significantly from the current 194 involved in UNFCCC negotiations to a Climate Council of around 12.[4] This proposal is specifically designed to uphold procedural legitimacy by including the most affected actors but also limiting participation in ways that will help secure agreement. Although Eckerlsey's minilateralism remains formally within the UNFCCC, it does so by excluding some states from key decision-making.

## Input and Output Legitimacy in International Negotiations

Superficially, perhaps, there are good theoretical reasons to think that a small group of states could reach a faster and more effective agreement than a broader group when consensus guidelines are in place.[5] At a basic level, each state has to take consideration of their own (domestic) preferences—their 'win-set'—in multilateral negotiations (Putnam 1988). As the number of negotiators increases, it becomes harder to find points of intersection where win-sets overlap and mutually agreeable outcomes can be reached between states. Turning toward minilateralism in an effort to secure a faster and more efficient agreement might then seem a logical step. It is worth emphasizing, however, that there is little concrete discussion in the literature on minilateralism about how a small group of states will reach faster outcomes. Indeed, if the differences which exist between states at the multilateral level are replicated in minilateral fora, then finding overlapping points of agreement becomes just as difficult.

This point aside, the core argument of this chapter is to resist exclusive and inclusive *minilateral* approaches (Victor 2011; Eckersley 2012). This rejection does not stem from procedural considerations grounded in fairness and equality (thought these are certainly important values). Rather, I argue that inclusive multilateralism is essential to secure the output legitimacy sought by minilateralists (Scharpf 1999). This is because output and input legitimacy are not part of a zero-sum game whereby decreasing input functions will increase output results (or vice-versa). Rather, recent work in democratic theory—especially the deliberative tradition—continues to highlight how inclusion and procedural fairness actually generate more effective decisions which facilitate collective problem-solving. Multilateralism and consensus deliberations do not just respect the autonomy of individuals but actually help to foster more effective decision-making. As such, inclusive multilateralism amongst a large number of states and TNAs is more likely to generate better, effective and efficient outcomes than smaller groups.

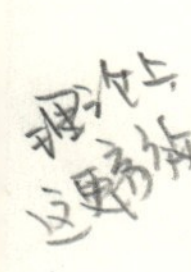

A minilateralist might contend that given the severity and unprecedented nature of climate change, we should seek a quick agreement and put our democratic (procedural) demands on hold. I maintain that this is a mistake. Tackling climate change requires *effective* policy and I elaborate why input is critical for this success. Because I recognize that time is of the essence, I suggest ways that multilateralism can be made more temporally sensitive while maintaining the inclusivity and procedural legitimacy. Given the global nature of climate change, then, increasing both input and output variants of legitimacy through inclusion is crucial as we think about how gridlock should be overcome.

This is an important argument because in academic and policy circles surrounding climate change governance, procedural inclusion is often thought to directly undercut the speed and efficiency of outcomes. Eckersley (2012, 25), for instance, states that these two positions—inclusive multilateralism and exclusive minilateralism—appear to be irreconcilable. Greg Hunt, the Australian Federal Minister for the Environment, similarly contends that the broad inclusion at Copenhagen directly undercut the likelihood of reaching (or even finding) a practical solution to advance negotiations.[6] It is also an important argument as we seek to assess and understand current minilateral attempts to tackle climate change. For instance, the Group of 8 (G8), the Asia-Pacific Partnership on Clean Development and Climate (APP) and the MEF have all failed to reach agreement. Within the G8, the issue of climate change has increasingly fallen off the radar since 2008, the APP concluded in 2011 with many projects incomplete and the MEF has failed to produce either short or long term targets for emission reduction (McGee 2011; Karlsson-Vinkhuyzen & McGee 2013). Moving forward, then, requires a reassessment of the relationship between input and output legitimacy for breaking gridlock.

## TWO ARGUMENTS FOR DEMOCRACY

Multilateral negotiations are supposed to be democratic in terms of input legitimacy—giving all relevant actors a fair say—and output legitimacy—facilitating the vigorous pursuit of the common interest and effective problem-solving that benefit everyone. On this score it is important to look at the intrinsic and instrumental rationalities embodied within democratic decision-making. Intrinsic reasons for following democracy relate to the importance of respecting all individuals as agents who are due consideration (for reasons of morality or legitimacy). Instrumental considerations, alternately, suggest that democracy is valuable as a decision-making method because it produces better outcomes than other methods of decision-making. It is these instrumental considerations, which help bridge input and output legitimacy, that require elaboration to show how and why minilateralism is deficient on its own terms. I discuss two instrumental ways in which inclusion induces better outcomes.

### Cognitive Diversity

The first democratic reason underpinning inclusive multilateralism can be derived from recent work on epistemic decision-making. The basic idea is that inclusiveness in the context of democratic deliberations helps produce smarter results than less inclusive debates (Landemore 2013). This argument, which finds resonance in the literatures on cognitive psychology and theoretical economics, stresses the importance of cognitive diversity when seeking practical solutions to different problems (Hong and Page 2004). From this stance, decision-making requires a diversity of perspectives (ways of representing situations), a diversity of interpretations (ways of categorizing perspectives), a diversity of heuristics (ways of generating solutions to problems) and a diversity of predictive models (ways of inferring cause and effect relationships) (Landemore 2013). Inclusivity across these dimensions helps generate outcomes which track independent truths and uncover mutually acceptable solutions.

The idea that diversity trumps ability—that a large selection of individuals outperforms a small collection of experts—has been demonstrated in a range of contexts. To give a fairly simple example from the area of cryptography: a broad group of individuals with disparate interests and diverse cognitive skills are more likely to crack a code than a small, homogenous expert group. In other words, a group with myriad alternative perspectives—from backgrounds such as mathematics, linguistics, music, natural sciences and so on—is more likely to crack a code than a group of computer programmers even if the latter group has much more advanced education and training. This is because the collection of experts only gains marginal increasing returns from the addition of new members, whereas the cognitively diverse group gains comparatively more for each new member (Hong and Page 2004, 16387-88; Landemore 2013, 1213). In this instance, large groups are better at solving problems when no initial answer is proposed as well as when selecting between prepackaged solutions.

In terms of climate change and the UNFCCC negotiations, the importance of cognitive diversity to collective problem-solving becomes obvious: minilateral groups may not contain a wide-variety of perspectives, interpretations, heuristics, or predictive models. For instance, the proposal by Victor and Names suggests limiting minilateral negotiations to a small club of high-emitters. The MEF—which Victor proposes as one way to begin minilateral discussions—comprises Australia, Brazil, Canada, China, the EU, France, Germany, India, Italy, Indonesia, Japan, Korea, Mexico, Russia, South Africa, the United Kingdom and the United States. While this group contains some variation in terms of level of development and world region, it is missing many other perspectives. Small island states, Africa, the Middle East and Central America are entirely absent and different political perspectives—such as communitarian Socialism (espoused by Bolivian President Evo Morales), feminism and indigenous rights—are missing. While Eckersley's position on inclusive multilateralism goes some way to help include different perspectives and interpretations, she seeks to limit the number of negotiation states by excluding middle income, median emitting and somewhat-affected states. This remains problematic from an epistemic position because these states may well have different and valid perspectives that should be brought out in climate change negotiations.

Recent work from Hayley Stevenson and John Dryzek (2012) has documented the types of perspectives (or in their words, 'discourses') offered by different actors in global environmental negotiations. The states involved in the MEF are, mostly, corporatist states that are economically powerful and high emitting. These states generally formulate climate policy within a paradigm of mainstream sustainability (Stevenson 2014, 182). This perspective accepts the existing political and economic infrastructure of world politics and generates solutions within those parameters. Alternately, middle income countries and developing states approach climate change from a perspective of expansive sustainability or green radicalism: the former accepts current economic incentives but seeks new political institutions whereas the latter demands systemic overthrow of existing authority and power distributions. Minilateral proposals offered by Victor and Eckersley seek to limit the number of participating states and thus the range of perspectives and heuristics in negotiations. Following the logic offered by Hong and Page, a smaller group with less diversity will be much less likely to generate effective and acceptable policy decisions.

Minilateralist proposals also entail reducing the number of TNAs involved in climate change deliberations. This move would also be problematic in terms of epistemic decision-making because non-state actors often voice perspectives ignored by states. The nine constituency groups at the UNFCCC include businesses, trade unions, farmers, women/gender, youth, research, local governments, indigenous and environmental TNAs.[7] These perspectives and their heuristics for generating responses to anthropogenic climate change vary widely. Nagmeh Nasiritousi and her co-authors (2014) have also documented the types of perspectives held by different non-state actors in the side-events of the UNFCCC process. They argue that public organizations, public-private partnerships, business lobbies, research groups and NGOs all hold different positions and voice different heuristics/predictive models to mitigate greenhouse gas emission. Technological innovation, market-based mechanisms, government regulation, low carbon economy and lifestyle changes are all advocated to different extents by different actors. While Nasiritousi et al. (2014, 182) are confident that non-state actors help pluralize the perspectives which infiltrate climate change politics, they suggest even more viewpoints are required to enhance the democratic standing of climate change negotiations. Minilateralism seeks to reduce the number and strength of different perspectives discussed in order to help foster agreement. Epistemically, this is highly problematic in terms of crafting intelligent policy.

Of course, many scholars have already argued that climate change requires a diverse range of viewpoints to be tackled effectively. In a variety of different books and articles, John Dryzek (1996) has contended that ecological problems cannot be mitigated through strategic rationality: Humans cannot simply decompile climate change into component issues and then seek to manipulate nature to serve our own ends. Rather, climate change should be treated as a complex whole and policy choices should be grounded in reflexive and inclusive deliberation. In light of Dryzek's claims, then, minilateralism approaches—which are exclusionary and technocratic—fare poorly. Because climate change is a broad issue that encompasses rising sea levels, melting ice caps, ocean acidification, global warming, desertification and myriad other interconnected problems, minilateralism risks removing perspectives that are necessary to forge *effective* policy as states grapple with alternatives for reducing carbon emissions.

A similar argument which links procedural inclusion to enhanced epistemic outputs can also be gleaned from the pragmatist tradition in democratic theory. Based on the writings of John Dewey, pragmatists stress the importance of cognitive diversity for collective problem-solving and critical intelligence. In this vein, social problems should be deliberated over by a wide group of individuals so that new cooperative solutions can be forged. In a recent article, Daniel Bray explicitly links inclusion in UNFCCC negotiations to enhanced epistemic quality of policy-making. Bray (2013, 466–68) claims that UNFCCC debates should continue to be based on the established weight of scientific evidence *and* local observations of impacts. As such, including local populations in climate change negotiations remains crucial to solve climate change issues. 'Communities around the globe', Bray (2013, 648) notes, 'have vastly different interpretations of the problem of climate change based on their particular experiences and values'. For the inhabitants of low-lying island countries, for instance, climate change is about surviving sea level rises, changing weather patterns and the loss of fresh water resources and fishing stocks. This varies considerably with the problems facing citizens of developed states related to climate change. The variegated local issues caused by climate change suggests that international negotiations should take consideration of these perspectives in order to create epistemically useful policy instead of deliberately limiting participants in the hope of reaching agreements faster.

Overall, then, there are good reasons to think that the epistemic quality of decision-making can be enhanced by including a variety of perspectives and heuristics in negotiations. The literature on democratic epistemology and especially the work by Landemore, suggests that a broad range of deliberators are more likely to find effective and appropriate solutions than a small group of experts. So while much more work needs to be done to say how international negotiations might be more deliberative, it is clear that inclusion is an essential condition. This argument helps show why the logic of minilateralism fails on its own terms: effective outputs (i.e., good policy) require inclusive inputs (procedures) so that alternative viewpoints are compared and tested against one another.[8] We thus have good reason to support open, inclusive and democratic multilateralism within the UNFCCC system.

## Consensus and Compliance

Producing epistemically desirable outcomes is just one instrumental benefit of inclusive multilateralism. Inclusive and democratic agreements are also more likely to secure compliance. This is especially vital because climate change is a collective action problem *writ large* (what some have described as a 'wicked problem' [Levin et al. 2012]).[9] Because the atmosphere is a common pool resource—making it both rivalrous and non-excludable—there is an incentive for some agents to free-ride on the actions taken by others. This is compounded by the corollary that mitigation (i.e., GHG reductions) is a public good that is both non-rivalrous and non-excludable. When thinking about international negotiations, then, we need to think about policies which mitigate the likelihood of free-riding and defection. The core claim of this section is that inclusive and democratic negotiations—that aim toward consensus—improve compliance. This argument is again drawn predominantly from the field of deliberative democracy.

Deliberative democrats have stressed the importance of argumentation and consensus between agents for ensuring compliance. Habermas (1993), for instance, grounds his theory of communicative action in the validity of speech acts. Essentially, agents who engage in non-coercive deliberation are attempting to understand one another. As opposed to strategic action (in which agents try to get others to do what they want), communicative action entails making validity claims which rest upon sincerity, rightness and truth. For Habermas, social order and cooperation depends upon the actors recognizing each other as agents, deliberating and reaching consensus. Deliberation, properly conceived, thus helps uncover mutually acceptable solutions to problems that agents are likely to accept.

The broader literature on deliberative democracy has followed Habermas and suggests that deliberative consensus helps secure compliance. John Parkinson (2003, 182) contends that non-coercive and reciprocal deliberation enables individuals to understand each other, accept reasons for action and ultimately generate consensus. This enhances the legitimacy of decision-making on both an intrinsic and an instrumental level. Intrinsically, it is morally desirable for agents to accept collective decisions and the exercise of authority if good reasons are provided. Instrumentally, Parkinson (2003, 182) contends that consensus between agents 'makes political processes more efficient by reducing the costs of enforcing compliance'. The fundamental logic is that agents who agree to a solution and agree on reasons for that action are internally motivated to act in line with that agreement.

Again, the link between consensus and compliance sheds light on how climate change negotiations should proceed. Minilateral agreements exclude states from deliberations that will ultimately need to comply with climate policy and reduce $CO_2$ emissions. Because climate change is a collective action problem with global scope, securing the compliance of a small group of states may not actually help reduce emissions. Excluded states would be tempted to free-ride on the policies of major emitters by increasing their emissions and businesses would be happy to off-shore production to those countries.[10] Inclusive multilateralism dampens this problem because states included in deliberation are more likely to comply with agreements that they helped negotiate.

Some recent literature has directly noted that minilateralism will undermine compliance with climate change policies. Ed Page (2011, 53) has contended that inclusion, impartiality and equal opportunity are essential to achieve environmental objectives. States are unlikely to comply with climate policies if 'large numbers of atmospheric users feel excluded from the policymaking process' (Page 2011, 54). Bray (2013, 465), in a similar claim, argues that inclusion and equality in the UNFCCC process is pragmatically desirable 'because its universal legitimacy in this context increases the political power of the treaty to secure widespread participation and compliance'.

Bray specifically turns this argument against minilateral approaches—such as that offered by Victor and Naim—by suggesting that the UNFCCC must remain the central avenue for overcoming climate negotiation gridlock. The fundamental point raised by deliberative democrats, such as Habermas and Parkinson, as well as climate scholars such as Page and Bray, is that inclusive deliberation is essential for securing compliance with policy and that compliance is necessary to tackle climate change effectively.

## BREAKING MULTILATERAL GRIDLOCK

In the previous section I argued that there are good reasons to think that inclusive deliberation is essential for reaching effective outcomes on climate change. This provides a powerful rejoinder to one horn of the minilateralist claim: that a small group of states will be able to secure an *effective* agreement to tackle climate change. However, the minilateralist claim also rests on the notion that the consensus-based UNFCCC is gridlocked and, after nearly two decades of impasse time is running out for meaningful policy to be created (Grasso and Roberts 2014). As such, it is vital that in advocating for an inclusive minilateralist argument on grounds of democratic instrumentality that I respond to the temporality dimension.

Some scholars have begun suggesting alternate ways to advance UNFCCC responses without resorting to minilateralism. Aaron Maltais (2014), for instance, suggests that a small group of states should take the lead in addressing climate change. This does not mean that a minilateral group should take it upon themselves to deal with climate change, but rather that a small group of economically powerful states should act unilaterally to demonstrate that domestic economic welfare is compatible with rapidly decreasing greenhouse gas emissions. This would provide the foundation for a more inclusive and comprehensive global climate policy. Alexander Ovodenko (2014), somewhat differently, shows that states have managed to move some issues outside the UNFCCC framework to deal with negotiation gridlock. Ovodenko explores how different institutional resources and sectoral participation mean that some topics—such as technology financing and emission mitigation—have been approached outside the UNFCCC, whereas other issues—such as forestry management[11]—have remained within UNFCCC purview. Ovodenko (2014, 175) argues that the overlapping institutions for governing climate change 'may relieve pressures to reform a single institution by providing governance opportunities when the demand for collective action grows'.

While these are certainly useful prescriptions, I turn toward another possible avenue for mitigating negotiation deadlock within inclusive multilateralism: the employment of flexibility mechanisms in UNFCCC agreements.

Specifically, I discuss how the current path-dependent arrangement of institutions limit response to climate change in multilateral fora and suggest how flexibility mechanisms can help maintain inclusive and democratic negotiations while making agreement more tractable.

## Path Dependence and Climate Governance

There is a growing body of literature that suggests that the current arrangement of global climate change architecture has developed path-dependently. Path dependence is a key concept in the field of economics and political science (specifically historical institutionalism) and suggests that institutions, once they are put in place, become 'locked in' over time. Through a variety of mechanisms—self-reinforcement, increasing returns and positive feedback—the cost of moving off an institutional 'path' rises as initial expenditures become sunk, actors learn new rules and norms of 'appropriate' action are redefined. The classic example in the literature comes from the emergence and persistence of the QWERTY keyboard. Although QWERTY is a sub-optimal keyboard layout for typing efficiency (compared to the Dvorak layout), this system has become locked-in over time through path dependence. The physical infrastructure of millions of QWERTY keyboards, the fact that individuals learn how to type on this layout and the norms associated with using major brands has reinforced the QWERTY system making wide scale change a virtual impossibility.

The global institutional system for mitigating climate change is often described as path-dependent. The UNFCCC (and Kyoto Protocol) is just one (central) aspect of global climate governance that also includes other multilateral deals (the Montreal Protocol), club initiatives (the MEF, APP, or the G8), subnational laws (such as the Californian emission trading system), a host of non-state regulatory efforts and private experiments (Keohane and Victor 2011, 10). This governance system—what Keohane and Victor refer to as the climate change regime complex—has emerged in piecemeal fashion as institutions for confronting climate change are introduced alongside one another. Because these institutions emerge organically, they tend to overlap in terms of mandate, issue-space and membership. Over the past 70 years this system has become locked-in as states and non-state actors have sunk costs in to these fora, learnt institutional rules and been socialized to accept the general framework for climate mitigation.

The path dependence of global climate architecture is actually embedded in a deeper set of path-dependent institutions. As John Dryzek (forthcoming) has noted, human civilization emerged during the Holocene epoch of the past 10,000 years. The Holocene is now giving way to the Anthropocene in which human actions drive changes in the Earth system. Unfortunately, several of the key institutions that developed late in the Holocene era—the nation-state, markets and global governance—were not designed to deal with climate change or its attendant problems. Given the stickiness of global climate institutions narrowly (Keohane and Victor 2011) and the broader framework of human institutions within which we must operate, it is crucial to think about how path dependence limits and creates space for new policy.

On the surface, the path-dependent nature of current climate change architecture might be cause for concern: the gridlocked UNFCCC system is locked-in place and becoming harder to change as time passes. However, this skepticism should be tempered. Although path dependence does make it harder to alter existing institutions, path dependence and positive feedback also provide opportunities to lock-in new and beneficial policies that are resilient. Moreover, a detailed understanding of how institutions foster path dependence also helps policy-makers identify possible entry points to enact institutional change.

Andrew Jordan and Elah Matt (2014) have suggested that path dependence is a useful explanatory tool for understanding how institutions develop and why they persist. They also contend that the concept holds promise for institutional designers in the area of climate change. Jordan and Matt analyze how the European Commission (EC) introduced Voluntary Agreements (VA) to push car manufacturers toward producing less polluting cars. In the beginning, this VA was fairly unsuccessful due to the path-dependent reasons: car manufacturers and major industries based on fossil fuels combined to oppose that VA. Eventually, in 1998, the EC produced a weak policy 'encouraging' change. Over time, this policy has been difficult to implement and has stalled on several occasions. Despite these setbacks, Jordan and Elah (2014, 232) contend that the EC has engaged in a policy of 'relentless incrementalism' designed to push manufacturers toward building low-emission vehicles. Regulation 443 in 2009—launched by the EC and the European Parliament—has seen success in this way as manufacturers were required to have 100% of their passenger car fleets emitting less than 130g/km of $CO_2$ by 2015.[12] Jordan and Matt use this example to show how policy-makers can overcome existing obstacles and lock-in policies that gain prominence over time.[13]

In a similar article, Kelly Levin and her co-writers (2012, 123) address how path dependence affects the 'super wicked' problem of climate change. Specifically, Levin et al. (2012) discuss how existing path-dependent institutions can be overcome and then how new policies can be introduced that take advantage of institutional lock-in to help solve climate change problems. This requires designing policy that becomes entrenched (reinforced) and expands (positive feedback) to cover wider groups of people. In the US context, these authors argue that gridlock at the national level meant that actors turned toward state-based initiatives to implement new policy. Because of the separation of powers, these initiatives became difficult to overturn and eventually became locked-in and diffused to other states. The Californian Cap-and-Trade Program and a variety of state initiatives to give consumers the ability to choose non-fossil fuel energy options on electricity bills demonstrate how existing problems can be overcome and new policies can be locked-in.

## Moving Forward: Flexibility Mechanisms and Policy Design

This recent literature suggests that, although the existing UNFCCC system is gridlocked, we should look for ways to overcome path-dependent problems and design new (beneficial) policies that take advantage of institutional lock-in and positive feedback. I follow this line of argumentation and suggest that flexibility mechanisms—which are often employed in international agreements—could be useful in the UNFCCC system to overcome gridlock in a timely manner. At their core, flexibility mechanisms offer a way to manage risks associated with joining an agreement by dictating the terms of cooperation. Generally speaking, flexibility allows actors to respond to uncertainty by altering institutional rules, changing the distribution of burdens and benefits and temporarily exiting an agreement in response to sub-optimal developments or exogenous shocks. This flexibility, in turn, makes it easier to reach agreement in the first place by expanding the win-sets of different states and shortening the time-horizons of negotiators (Fearon 1998).[14]

In response to multilateral gridlock, then, flexibility mechanisms offer a way to help reach agreement. The key is to find and employ flexibility in design that helps balance change against stability. Agreements must be flexible enough so parties sign-on but provide enough stability so the agreement persists in the future (Fearon 1998). The question then becomes: which flexibility mechanisms might help the UNFCCC overcome gridlock but not provide so much flexibility that agreements are either discarded or diluted over time?

Turning to the existing climate change system proves instructive. The UNFCCC currently offers several mechanisms for enhanced flexibility. The generalized nature of the 'framework protocol'—which outlines overarching principles, objectives and institutional rules—provides some flexibility. As we have seen, the framework protocol has spawned the 1985 Vienna Convention for the Protection of the Ozone Layer, the 1987 Montreal Protocol and the Kyoto Protocol. Moreover, each COP works within the general UNFCCC and decisions taken at each COP does not amend the framework-protocol approach.[15] The UNFCCC system also contains a 'review and amendment' rule. Article 7.2(i) of the UNFCCC stipulates that states should periodically re-assess the obligations of each party and the institutional structure of the Convention in light of previous experiences and new scientific knowledge.[16] The Subsidiarity Body for Scientific and Technological Advice is in place to facilitate this process.

The Kyoto Protocol, adopted in 1997, employed several different flexibility mechanisms to help states reach agreement and begin the process of ratification (Guldbransen and Andresen 2003). The Clean Development Mechanism, for instance, enables states to implement an emission-reduction project in developing countries and earn saleable certified emission reduction credits. Emission trading, as a market-based mechanism, gave Kyoto signatories more flexibility. Annex I states that emit less than their quota can sell emissions to states that exceed their quota. Perhaps most notably, the Kyoto Protocol also allowed withdrawal for signatory states within five years of ratification. Canada, in late 2011, adopted this strategy and withdrew from the Kyoto Protocol from 2012 onward, thus avoiding financial penalties associated with excess emissions. The shift toward Intended Nationally Determined Contributions (INDCs) in post-Warsaw discussions also reflects a new, flexible approach to climate change commitments. Despite the wide-range of flexibility mechanisms built in to the UNFCCC generally—and the Kyoto Protocol specifically—it remains clear that the existing regime struggles to facilitate agreement and reduce emissions (Ovodenko 2014). Although the state of current research does not show us exactly how new flexibility mechanisms could be employed to overcome cooperative gridlock, some general suggestions can be gleaned from the preceding comments and previous work.

I suggest that escape clauses offer a particularly useful device for tackling climate change. This is an argument I have defended elsewhere on grounds of democracy and feasibility (see respectively: Kuyper 2013; Kuyper forthcoming). Escape clauses allow states to derogate temporarily from an agreement in the event of a major unexpected shock, while assuring other negotiating parties a return to compliance after escape (Pelc 2009). When employing an escape clause, typically the derogating state has to pay some form of compensation to others. This mechanism is used in many different international organizations such as the bilateral investment deals, the World Intellectual Property Organization (WIPO) and, perhaps mostly notably, the World Trade Organization (WTO). Of course, many other flexibility mechanisms—such as withdrawal clauses, sunset provisions, amendment rules and reservation principles—can also be employed to help states overcome bargaining issues and secure agreement (Marcoux 2009). And indeed, the appropriate mechanism (or mix thereof) required for overcoming UNFCCC gridlock is a matter for future investigation. I do provide reasons below, however, to show how and why escape clauses might be theoretically and empirically appropriate for the problem at hand.

Escape clauses help balance change and stability in useful ways: states can make an initial agreement with the knowledge that rules can be temporarily set aside in extreme cases. As such, escape clauses give states flexibility that, even when employed, does not break the initial agreement. This responds well to the issues of path dependence discussed previously. Escape clauses make it easier to overcome the current stickiness within the climate change negotiations by giving states the ability to respond to unforeseen future outcomes, thus making agreement more likely. These mechanisms also take advantage of institutional lock-in by ensuring that the agreement persists even when faced with disagreement (Levin et al. 2012).

Here it might be useful to provide a brief case description to help make this point more tangible and show how the deployment of flexibility mechanisms helps allay minilateralist concerns over current gridlock. As a parallel example, escape clauses are embedded in the WTO, such as article XIX of the General Agreement on Tariff and Trade (GATT) and the Agreement on Safeguards. The institutional codification of escape clauses in the GATT framework was essential for fostering initial agreement and ratification between states. Escape clauses are used by states to defect temporarily from agreed-upon production and tariff levels to protect domestic markets when a severe and exogenous shock hits. During the 1950s, escape clauses were activated regularly in the GATT (Pelc 2009). Over time, however, their usage has become less frequent as states rely on 'appeals to exception' when breaching formal rules. Appeals to exception work as the derogating state provides justificatory reasons for why they cannot remain compliant within formal rules for a set period of time due to an issue outside of their control. Other states (and the Appellate Body) can accept or reject these appeals and generally need information about the severity and exogeneity of a shock to determine whether an appeal is worth granting.

The significant numerical reductions in activation of escape clauses provide reason to think that, over time, states can recognize that employing escape mechanisms (and paying both reputational and monetary costs) is not worthwhile. Although all states have an incentive to have escape mechanisms available for usage, they also have an incentive for a system that allows 'appeals to exception' so that escape does not need to be invoked. Escape clauses thus provide states with flexibility needed to overcome gridlock. Over time, appeals to exception can be used so that escape is highly infrequent. Given the power and persistence of the GATT/WTO, this shows how escape clauses can be used early in agreements to kick-start an institution while other mechanisms can arise gradually to help ongoing institutional functioning and cooperation. Outside of the GATT/WTO, escape clauses have also been vital in aspects of the climate change regime, such as the London Convention dumping regime, the Ramsar regime on wetlands, the Tropical Timber regime and parts of the Oil Pollution regime (Marcoux 2009, 220).[17]

Taking these arguments back to the UNFCCC it becomes clear that different forms of flexibility are yet to be tried in the climate change context. Although withdrawal clauses were written in the Kyoto Protocol, the complete exit of Canada in 2011 exposed the problem with that mechanism. Instead, escape clauses might be useful in forging initial agreement between states while enabling flexibility in continuing cooperative relationships. Indeed, Scott Barrett (2003) has directly noted that the lack of escape clauses in the Kyoto Protocol was a major misstep in institutional design. Because flexibility mechanisms provide an alternate way to help foster agreement between dissenting states, this addresses the minilateralist concern about temporality. Escape clauses may allow a wide-range of state and non-state actors to participate in climate negotiations which enhances both epistemic decision-making and post-agreement compliance. This suggests that flexible, inclusive and transparent processes may actually be compatible with producing effective and efficient outcomes in a timely manner.

## SUMMARY AND CONCLUSION

I have argued that minilateral proposals to break gridlock in UNFCCC negotiations are flawed. Although there are certainly good reasons in terms of legitimacy to support inclusive multilateralism, I have suggested that there are also good instrumental reasons. Specifically, inclusivity generates epistemic benefits and enhances the likelihood of compliance. Given the nature of climate change, both intelligent problem-solving and collective action are essential for meaningful policy-making. This argument further suggests input and output forms of legitimacy are not necessarily in tension and can, in some circumstances, be mutually supportive. I have also contended that employing certain flexibility mechanisms may help overcome gridlock in climate change governance and lock-in beneficial agreements. Although this argument needs more sustained analysis, this should certainly stand as a future direction for promising research.

There remain several other important lines of research. Although inclusivity is advantageous for reasons of improved cognition and compliance, much of this benefit hinges upon the quality of deliberation between actors. Ideally, deliberators should provide good reasons for their position, listen to others and weigh competing counter-claims when forming their opinions and preferences.[18] During international negotiations, this standard is demanding as national interests and *realpolitik* come to the fore. So while inclusion is a prerequisite for good outputs, much attention should be paid to the quality of deliberation between parties.

Related to this, normative and empirical scholars should focus on how TNAs can enhance the viewpoints offered in climate change negotiations both within the UNFCCC system and the wider climate change regime. Recent research has systematically shown that international organizations are increasingly ‘opening up’ to TNAs in different ways (Tallberg et al. 2013).

For example, the Governing Council of the United Nations Environment Programme, the Environmental Ministerial Meeting of the Association of Southeast Asian Nations (ASEAN) and the Environmental Policy Committee for the Organization for Economic Co-Operation and Development (OECD) have all increased access for TNAs over the past 35 years. This increased inclusion holds much promise for the epistemic quality of international negotiations which should be harnessed rather than undercut (as minilateralists are wont to do).

Overall, minilateral prescriptions fail to provide a suitable approach to the issue of climate change. Actual minilateral efforts (G8, APP, or the MEF) have failed in their goals because the key actors involved have sharply polarized views on climate governance. Instead of continuing down the minilateral path, we should recognize that inclusivity comes with attendant benefits. Drawing upon deliberative democracy and cognitive psychology, I have elaborated more precisely how inclusivity generates more effective outcomes. So while gridlock must be overcome—especially as we approach COP 21 in Paris at the end of 2015—paying more attention to the process of UNFCCC negotiations could help improve outputs.

## REFERENCES

Abbott, Kenneth W. 2012. 'The Transnational Regime Complex for Climate Change'. *Environment and Planning C: Government and Policy* 30(4): 571–90.

Barrett, Scott. 2003. *Environment and Statecraft: The Strategy of Environmental Treaty-Making*. Oxford: Oxford University Press.

Bray, Daniel. 2013. 'Pragmatic Ethics and the Will to Believe in Cosmopolitanism'. *International Theory* 5(3): 446–76.

Bäckstrand, Karin. 2008. 'Accountability of Networked Climate Governance: The Rise of Transnational Climate Partnerships'. *Global Environmental Politics* 8(3): 74–102.

Depledge, Joanna, and Pamela Chasek. 2012. 'Raising the Tempo: The Escalating Pace and Intensity of Environmental Negotiations'. In Pamela Chasek and Lynn Wagner (eds.), *The Roads from Rio: Lessons Learned from Twenty Years of Multilateral Environmental Negotiations* (New York and London: Routledge), 19–38.

Dryzek, John S. 1996. 'Foundations for Environmental Political Economy: The Search for Homo Ecologicus'. *New Policy Economy* 1(1): 27–40.

———. Forthcoming. 'Institutions for the Anthropocene: Governance in a Changing Earth System'. *British Journal of Political Science.* DOI: http://dx.doi.org/10.1017/S0007123414000453.

Eckersley, Robyn. 2012. 'Moving forward in the climate negotiations: multilateralism or minilateralism?' *Global Environmental Politics* 12(2): 24–42.

Fearon, James. 1998. 'Bargaining, Enforcement and International Cooperation'. *International Organization* 52(2): 269–305.

Grasso, Marco, and J. Timmons Roberts. 2014. 'A Compromise to Break the Climate Impasse'. *Nature Climate Change* 4(7): 543–49.

Guldbransen, Lars H., and Steinar Andresen. 2003. 'NGO Influence in the Implementation of the Kyoto Protocol: Compliance, Flexibility Mechanisms and Sinks'. *Global Environmental Politics* 4(4): 54–75.

Habermas, Jürgen. 1993. *Justification and Application: Remarks on Discourse Ethics*. Translated by Ciaran Cronin. Cambridge: Polity.

Hong, Lu, and Scott E. Page. 2004. 'Groups of diverse problem solvers can outperform groups of high-ability problem solvers'. *Proceedings of the National Academy of Science* 101(46): 16385–89.

Jamieson, Dale. 2014. *Reason in a Dark Time*. Oxford: Oxford University Press.

Jordan, Andrew, and Elah Matt. 2014. 'Designing policies that intentionally stick: policy feedback in a changing climate'. *Policy Sciences* 47(3): 227–47.

Karlsson-Vinkhuyzen, Sylvia, and Jeffrey McGee. 2013. 'Legitimacy in an Era of Fragmentation: The Case of Global Climate Governance'. *Global Environmental Politics* 13(3): 56–78.

Keohane, Robert O,. and David G. Victor. 2011. 'The regime complex for climate change'. *Perspectives on Politics* 9(1): 7–23.

Krebs, Ronald R. and Aaron Rapport. 2012. 'International Relations and the Psychology of Time Horizons'. *International Studies Quarterly* 56(3): 530–43.

Kuyper, Jonathan W. 2013. 'Designing Institutions for Global Democracy: Flexibility through Escape Clauses and Sunset Provisions'. *Ethics and Global Politics* 6(4): 195–215.

———. Forthcoming. 'Advancing Justice and Democracy "Beyond" the State: Feasibility through Flexibility'. *Political Studies*. DOI: 10.1111/1467-9248.12186.

Landemore, Hélène. 2013. 'Deliberation, cognitive diversity and democratic inclusiveness: an epistemic argument for the random selection of representatives'. *Synthese* 190(7): 1209–31.

Levin, Kelly, Benjamin Cashore, Steven Bernstein, and Graeme Auld. 2012. 'Overcoming the Tragedy of super wicked problems: constraining our future selves to ameliorate global climate change'. *Policy Sciences* 45(2): 123–52.

Maltais, Aaron. 2014. 'Failing International Climate Politics and the Fairness of Going First'. *Political Studies* 62(3): 618–33.

Marcoux, Christopher. 2009. 'Institutional Flexibility in the Design of Multilateral Environmental Agreements'. *Conflict Management and Peace Science* 26(2): 209–28.

McGee, Jeffrey. 2011. 'Exclusive minilateralism: an emerging discourse within international climate change governance?' *Portal Journal of Multidisciplinary International Studies* 8(3):1–29.

Müller, Patrick and Peter Slominski. 2013. 'Agree Now—Pay Later: Escaping the Joint Decision Trap in the Evolution of the EU Emission Trading System'. *Journal of European Public Policy* 20(10): 1425–42.

Naím, Moisés. 2009. 'Minilateralism: The Magic Number to Get Real International Action'. *Foreign Policy* 173: 135–36.

Nasiritousi, Nagmeh, Mattias Hjerpe, and Katarina Buhr. 2014. 'Pluralising climate change solutions? Views held and voiced by participants at the international climate change negotiations'. *Ecological Economics* 105: 177–84.

Ovodenko, Alexander. 2014. 'The Global Climate Regime: Explaining Lagging Reform'. *Review of Policy Research* 31(3): 173–98.

Page, Edward A. 2011. 'Cosmopolitanism, Climate Change and Greenhouse Emissions Trading'. *International Theory* 3(1): 37–69.

Parkinson, John. 2003. 'Legitimacy Problems in Deliberative Democracy'. *Political Studies* 51(1): 180–96.

Pelc, Krzysztof. 2009. 'Seeking Escape: The Use of Escape Clauses in International Trade Agreements'. *International Studies Quarterly* 53(2): 349–68.

Putnam, Robert D. 1988. 'Diplomacy and Domestic Politics: The Logic of Two-Level Games. *International Organization* 42(3): 427–61.

Ruggie, John G. 1992. 'Multilateralism: The Anatomy of an Institution'. *International Organization* 46 (3): 561–98.

Scharpf, Fritz W. 1999. *Governing in Europe: Effective and Democratic?* Oxford: Oxford University Press.

Stevenson, Hayley. 2014. 'Representing Green Radicalism: The Limits of State-based Representation in Global Climate Governance'. *Review of International Studies* 40(1): 177–201.

Stevenson, Hayley, and John S. Dryzek. 2012. 'The legitimacy of multilateral climate governance: a deliberative democratic approach'. *Critical Policy Studies* 6(1): 1–18.

Tallberg, Jonas, Thomas Sommerer, Theresa Squatrito, and Christer Jönsson. 2013. *Opening Up: International Organizations and Transnational Actors*. Cambridge: Cambridge University Press.
Victor, David G. 2009. 'Plan B for Copenhagen'. *Nature* 461(7262): 342–44.
———. 2011. *Global warming gridlock: creating more effective strategies for protecting the planet*. Cambridge: Cambridge University Press.
Zelli, Fariborz. 2011. 'The fragmentation of the global climate governance architecture'. *WIREs Climate Change* 2(2):255–270.

## NOTES

1. This research was made possible by grants from Riksbanken Jubileumsfond and the European Research Council (200971 DII) as well as the Swedish Research Council (Project No. 421-2011-1862) and Formas (Project No. 2011-779).

2. In this sense, multilateral negotiations are supposed to uphold several key tenets of democratic legitimacy. I further specify and develop this point as the chapter progresses.

3. There is an alternate view that Copenhagen was unexpectedly successful as states gave up on the notion of a binding treaty and moved toward the implementation of more realistic nationally determined targets. However, minilateralists and multilateralists often point at Copenhagen as a failure.

4. This number includes the USA, the EU, Japan, Russia, China, India, Brazil, South Korea, Mexico, the alliance of small island states (AOSIS), the African Group, and a bloc of LDCs.

5. To be clear, the UNFCCC draft rule 42 stipulates that matters 'of substance' should be adjudicated by consensus. Only if the consensus effort fails can a two-thirds or three-quarters majority be employed. In practice, deviation from consensus is rare (Yamin and Depledge 2004).

6. Greg Hunt, 'After Copenhagen: Time for the Major Economies Forum'. Available at: http://greghunt.com.au/Media/OpinionPieces/tabid/88/articleType/ArticleView/articleId/1339/Opinion-piece-published-in-The-Australian--After-Copenhagen-Time-for-the-Major-Economies-Forum.aspx (accessed 17/10/2014).

7. Non-governmental organization constituencies at the UNFCCC. Available at: http://unfccc.int/files/parties_and_observers/ngo/application/pdf/constituencies_and_you.pdf (accessed 14/10/2014).

8. Again, to reiterate, the minilateralist position rests on the dual pillars of effectiveness and temporality: we require a quick decision that helps mitigate climate change. My argument is mostly directed toward the effectiveness dimension (i.e., minilateralists want a policy that can actually help reduce the harmful effects of climate change). A minilateralist may still appeal to reduced input on the grounds of speed, but a quick agreement that fails to tackle the issue is surely not productive. Moreover, the minilateralist literature remains sparse on how a small group of high emitting countries who tend to have polarized views on who is responsible for emission reduction (between, say, China, India, and the US), would actually reach a faster decision. Because I recognize the importance of speed, however, I tackle this issue directly in the fourth section.

9. Super wicked problems are characterized by four features: time is running out; those who cause the problem are required to fix it; no central authority exists to address the problem; policy responses generally discount the future irrationally.

10. Indeed, we have already seen the EU engage in the off-shoring of carbon intensive productions (Grasso and Roberts 2014, 547).

11. Most notable is the program on Reducing Emissions from Deforestation and Forest Degradation (REDD+).

12. See 'Reduction in $CO_2$ emissions of new passenger cars'. Available at: http://europa.eu/legislation_summaries/internal_market/single_market_for_goods/motor_vehicles/interactions_industry_policies/mi0046_en.htm (accessed 14/10/2014).

13. For a similar argument, see Müller and Slominski (2013). These authors argue that flexibility mechanisms help foster green cooperation by spreading the costs of agreement over time.

14. The shortening of time horizons is a key mechanism underpinning the use of flexibility mechanisms. Short time horizons make it easier for negotiators to accept an initial agreement with some uncertainty because the institutional effects (equilibrium) will not last for a long time and thus potential costs are reduced. Interestingly, shortening time horizons is also psychologically important for effective agreements because individuals have been shown to make more realistic and feasible assessments when thinking about short-term implications rather than long-term problems or promises (Krebs and Rapport 2012).

15. Of course, this flexibility in framework-protocol is balanced against the tacitly accepted consensus rule.

16. 'United Nations Framework Convention on Climate Change Handbook'. Available at: http://unfccc.int/resource/docs/publications/handbook.pdf (accessed 21/10/2014).

17. This simply helps show that the GATT/WTO might be generalizable to other institutional issues within the issue area of environmental management. More explicit research on this is necessary, however.

18. 'Good reasons', in a deliberative sense, are bound up with the notion of 'public reasons'. In this, rules should be justifiable and acceptable to all parties over whom those rules have authority.

*III*

# Motivating the Present to Act for the Future

*Chapter Five*

# Making Our Children Pay for Mitigation[1]

Aaron Maltais

## INTRODUCTION

Investments in mitigating climate change have their greatest environmental impact over the long term. As a consequence the incentives to invest in cutting greenhouse gas emissions today appear to be weak. In response to this challenge, there has been increasing attention given to the idea that current generations can be motivated to start financing mitigation at much higher levels *today* by shifting these costs to the future through national debt. Shifting costs to the future in this way benefits future generations by breaking existing patterns of delaying large-scale investment in low-carbon energy and efficiency. As we will see in this chapter, it does appear to be technically feasible to transfer the costs of investments made today to the future in such a way that people alive today do not incur any net cost (e.g., Foley 2009; Rendall 2011; Broome 2012; Rezai et al. 2012; Rozenberg et al. 2013). The basic idea then is that governments can break current patterns of delaying mitigation investments by ensuring that their existing constituents do not need to make significant sacrifices.

The normative argument that we should finance mitigation by 'borrowing from the future' can be advanced in two general ways. The first is based on the empirical prediction that we will continue to see a pattern of very weak motivation among current generations to accept short-term mitigation costs. Thus, unless it becomes economically beneficial over the short-term to *markedly* increase investments in low-carbon energy and efficiency we should not expect to see sufficient investment to avoid dangerous levels of global warming. In this view, finding a way to pass on the costs of mitigation to future generations is an imperfect solution to the problem of weak moral motivation today but much better than the status-quo (Broome 2012, 37–48). In the second view, because we have good reason to expect that people in the future will be wealthier than people today (at least over the next century or so) and because the benefits of mitigation largely benefit people in the future, passing on most of the costs of mitigation to the future is actually a fair way to distribute these costs (Rendall 2011). Notice that the second view is not dependent on the empirical premise that people today will not be motivated to make sufficient short-term sacrifices, although the problem of motivating the present will give additional support to the argument for redistributing costs to the future.

In this chapter I focus on the implications of the first approach. Specifically, the aim of this chapter is to take seriously the possibility that climate change has produced an extremely intractable political problem and that we must now consider strong measures that can break existing patterns of delaying mitigation. I defend the claim that if climate change involves a stark conflict of interests between current and future generations, then borrowing from the future would be both strategically and normatively much better than the status quo.[2] However, I nevertheless challenge the borrowing from the future proposal on the grounds that it is not in fact the powerful tool for motivating existing agents that its proponents imagine it to be. The purpose of developing this critical argument is not, however, simply to throw doubt onto the idea of borrowing from the future.

Debt financing climate mitigation is a form of intergenerational buck-passing. In the climate ethics literature this type of buck-passing is usually viewed as deeply objectionable. As a consequence, normative theorising about climate governance tends to focus on institutional reforms that better represent the interests of future generations and inhibit buck-passing. My ultimate concern in this chapter is to argue that we cannot limit prescriptive normative theorising about climate governance to these types of reforms. If we really do find ourselves in a political context where the prospects for effective action are very poor then strategic forms of buck-passing may also make an important positive contribution to avoiding dangerous global climate change. Consequently, if debt financing is not as powerful of a motivational tool as imagined we still have strong reasons, I will argue, to identify other strategies that will change agents' incentive structures. To this end, I propose an alternative form of passing on the costs of mitigation to the future that warrants consideration.

The chapter is organised into five sections. Section I accounts for why motivating existing agents to invest in climate mitigation is taken to be such a difficult challenge. Section II defends the view that borrowing from the future can be normatively justifiable. Section III explains how it is thought to be possible to dedicate significantly more resources to mitigation today without current agents experiencing this as a cost. Section IV challenges the idea that borrowing from the future is a powerful tool for motivating the present to invest in mitigation. Section V proposes that we consider the development of an alternative form of explicitly pre-committing the future to mitigation costs. I defend this type of governance instrument at a normative level, specifically against the objections that it is 1) a form of tyranny of the present over the future and 2) morally corrupt.

## THE WICKEDNESS OF TIME IN THE ANTHROPOCENE

The capacity of the atmosphere, the oceans, and other natural sinks to safely carry GHG emissions is a global common pool resource. The mitigation of climate change is a global public good. We currently find ourselves in familiar conditions of unsustainable use of common resources and under provision of public goods. However, these cooperative challenges appear to be unusually difficult in the case of global warming. This is in part because of the large number and variety of actors that need to be coordinated and because of the scale and influence of special interests in the fossil fuel sector. In addition, there is some divergence between where the impacts of climate change will be most severe and which countries must bear the greatest mitigation costs. We are also currently lacking a leading state with strong incentives to act unilaterally and to coordinate other states. The high cost, complexity, and technological uncertainty involved in reforming our economies is another commonly highlighted obstacle. Yet, of all these impeding factors it is the role of time that appears to be the most toxic feature of this political problem.

In a recent paper, Hansen et al. (2011) estimate that it takes 100 years to see 60–90 percent of the warming response associated with GHG emissions. Long time lags in the climate system between emissions and temperature responses, between temperature stresses and damaging environmental consequences, and between effective mitigation policies and substantial infrastructure reform create a situation where investments in mitigation make little difference to the climate damages agents will experience over their lifetimes. It is past and probable on-going emissions that will have the greatest effects on current generations. From a game theoretic perspective this means that the relevant agents *do not* share in a preference for the collectively cooperative outcome compared to the collectively non-cooperative outcome. The consumption interests of people alive today are best promoted by *not* mitigating climate change.[3] In a more typical common problem, it is the agents' preferences for the collectively cooperative outcome that can be leveraged to establish norms and institutions that allow individuals to escape prisoner's dilemma dynamics (e.g., Ostrom 1990). In the climate case there is no prisoner's dilemma between generations. Rather, the central problem is to motivate existing agents to invest in protecting the commons for future agents (Gardiner 2001).

As time moves forward and irrespective of the climate conditions each generation is born into the same problem of weak incentives will be present (Gardiner 2001). Overcoming the intergenerational structure of the problem requires that agents be motivated by the interests of others (i.e., future agents) to a much greater degree than is true for agents in typical collective action problems. Norms of fair reciprocity may simply not be enough. Importantly, if we can redress the intergenerational motivational obstacle we can still face a typical global public goods problem between states. Even more importantly, as time passes the costs of mitigation increase, environmental damages increase, and the amount that has to be invested into adapting to climate change increases (Vaughan et al. 2009; Luderer et al. 2012; Rogelj et al. 2013). This means that the passing of time has the real potential to create a positive feedback where delay breeds stronger and stronger incentives for further delay (Shue 2010; Gardiner 2011, 185–209). The assessment above is not, of course, an attempt to give a full account of individuals' or political communities' motivations or to depict how individuals and groups have actually responded to climate change. The account above aims only to describe key obstacles to effective climate politics that can help us understand why the world's states have yet to invest in mitigation in a way that responds to the seriousness of the threat and why the politics of climate change appear so intractable.

## GIVING THE FUTURE A CHANCE TO PAY

If the incentives for passing on the costs of climate change to future generations are very strong, one response is to try to identify ways of passing on these costs in a way that best serves the interests of future generations. This is the core idea behind debt financing of climate change mitigation. Let us assume for now that we can pass on the costs of mitigation to the future in a way that does produce an improvement for these future agents compared to business as usual. Our terms of negotiation with the future are only possible because we are in a position of domination over them. We are free to ignore the fact that continuing to pollute the atmosphere will undermine the climatic conditions for human well-being far into the future. Thus, it appears to be disingenuous to claim that we are somehow helping the future by letting them pay for mitigation when it is our actions that are putting them in danger in the first place. This assessment has strong normative force, but there are also strong strategic and normative arguments for borrowing from the future.

To the extent that we expect political inertia to continue or worsen, identifying a no-cost option that could bring about immediate and significant mitigation investment while at the same time improving conditions for all the relevant agents leads to a very good outcome compared to perpetual delay (Broome 2012, 43–48). However, this improvement on the status quo is hardly a second-best option. Finding ways to bring the interests of present people and future people into better alignment, or finding ways to better mobilise the concern for future people by the current generation already appear to be much more normatively attractive options. Thus, we have good reason to be critical of failures to engage seriously in efforts to spread more climate friendly preferences or to make the long-term consequences of public policy more salient in political discourse. At the same time, if climate change is the most difficult cooperative challenge humanity has ever faced this difficulty must make some difference to our moral assessments of current failures to act and of strategies to address these failures.

The development of highly productive economies driven by the exploitation of cheap and abundant energy has been one of the main drivers of the amazing improvements in human welfare over the past two centuries. Individuals, companies, and governments both have had and continue to have good reasons for using fossil fuels. At the same time, finding ways to transition to low-carbon economies is straining the capacity of our economic and political institutions. Investing in the interests of the present has traditionally and continues to pass on enormous benefits to future people. At the same time, when it becomes clear that investing in shared goods today can undermine human welfare far off into the future it also becomes clear that we are straining the ability of our systems of morality to continually improve on the human condition.

The point is not to deny that it is deeply problematic that we are failing to dedicate a small fraction of current wealth to protecting the conditions for human welfare for generations to come. However, we must also acknowledge that climate change is a system level problem similar to other system-level problems in capitalist economies that are not intended and for which it is difficult to assign moral responsibility. From this perspective, taking on long-term debt to finance low-carbon infrastructure for the future is in part a moral failure but also in part a system-level response to a system level problem, similar to the way in which deficit spending to redress the effects of boom-bust cycles is a system-level response to the vulnerabilities capitalist economies generate.

We really do find ourselves in conditions of political delay with no sense of how or when these patterns might be broken. As a result, there is a strategic and normative case for at least some significant borrowing from the future. In fact, the proposal raises the following question: If we can solve the largest environmental threat to human welfare without anybody having to give up anything, then why don't we? Is it plausible to think that it is because we have simply failed to notice the options available to us? In the following section I explain how borrowing from the future to finance climate mitigation is technically possible. However, in section IV I argue that it is not surprising that we have not yet used debt financing to pay for mitigation. This is because the borrowing from the future proposal does not adequately address how costly it would be to compensate the current generation to the no-sacrifice level.

## WHAT DOES THE FUTURE HAVE TO BARGAIN WITH?

If we only have access to resources in the present, how can we direct these resources towards mitigation without this being perceived as a cost today? To achieve this the basic idea is that we can 1) change the *composition* of the savings we make for the future and 2) change the *composition* of the consumption bundles we will enjoy over our lifetimes (Broome 2012, 37–48). These changes in how we save and what we consume can, it is argued, free up resources for mitigation investments but at the same time involve no net cost. To see how this is expected to work, we can first look at changes in the way we save.

Each generation passes on savings to the next generation by leaving natural resources and by investing in things like infrastructure, technology, and knowledge that pass on productive capacity to future generations. However, because the true social costs of GHG intensive consumption and investment choices are not internalised it is argued that the current generation is actually saving for future generations in a very inefficient way. We could save for the future in a much more efficient way by shifting some current investment away from conventional capital and into mitigation capital, i.e., low-carbon energy, low-carbon infrastructures, and efficiency. By investing the resources necessary to avoid dangerous levels of global warming much more welfare is 'passed on' to future generations in the form of avoided climate damages than would be passed on to them in the form of conventional productive capacity. The idea is that a shift in the *composition* of the current generation's future-oriented investments can leave consumption levels constant. This brings us to the second issue: changes in the composition of our consumption.

The aim is to bring about an intergenerationally optimal level of investment in mitigation capital and an intergenerationally optimal shift away from GHG intensive consumption without affecting (too much) the value of lifetime consumption bundles for present people. In economic theory this is ideally achieved by the imposition of an optimal cost for GHG emissions. Compensating the present for making the consumption of GHG intensive goods more expensive can be achieved, it is argued, by consuming more goods that are not GHG intensive. Eating meat and other animal-based foods can be compensated with eating less expensive and higher quality vegetable based foods. Travel by car can be compensated with increased investment in public transportation. Buying less carbon-intensive consumer products and going on few overseas vacations can be compensated by consuming more services and working less.

Of course, the substitutions noted above are all already available to us and are not currently chosen to nearly a sufficient extent. The mitigation without sacrifice proposal cannot simply be that current agents should change their preferences. This is not because there are necessarily few opportunities for existing agents to change their preferences. Instead, the borrowing from the future proposal aims to show that even if we depart from the pessimistic premises that 1) existing agents are only willing to make modest sacrifices for the future and 2) that we can only expect existing agents to alter their preferences marginally, it is still possible to compensate these agents for investing in avoiding future climate damage. As a consequence, the argument has to be that there is some increase in alternative consumption patterns that the current generation prefers more than or at least as strongly as its current emissions-intensive economy.

If we impose an intergenerationally optimal carbon tax the costs of consuming emissions-intensive goods are increased and the returns on emissions intensive investments are decreased. The results are reductions in lifetime consumption as a direct response to cost increases and reductions in consumption as consequence of reductions in the rate of economic growth *over existing agents' lifetimes* compared to a business as usual (BAU) investment scenario. We can in part compensate for these losses by redistributing emissions taxes back to citizens and in part by taking on national debt (Foley 2009). The aim of this borrowing is to give the current generation enough of an alternative lifetime consumption bundle to make it worth its while to accept the effects of the carbon tax. To illustrate how this debt financing is expected to amount to borrowing from the future we can set up this borrowing via a pay-as-you go pension system.

Let us say that in the current pension system workers pay for retirees' pensions by transferring 5 percent of their earnings (i.e., productivity) into the scheme. Workers are motivated to make such transfers because they expect their children to pay for their pensions when they themselves retire. This allows workers to spread out their consumption between their productive and non-productive years and to save in a way that gives them access to some of the gains of economic growth in the economy. This same reasoning will hold for the worker's children and so on. This system of saving is a form of indirect reciprocity where the working generation confers a benefit on the retired generation in exchange for a future benefit from the young generation. This system of reciprocity does not rely on altruism, does not have a determinate endpoint, and users expect it to reach far out into the future (Heath

2013). To borrow from the future workers are asked to dedicate an additional 1 percent of their productivity to investment in mitigation capital. As compensation the workers' children will increase the size of the transfers they make to retirees when they themselves become workers. Our children will in turn be compensated when they retire by their children. When the benefits of avoided climate change begin to arrive workers can begin to reduce the amount of compensation they give to retirees.

Retirees are now being compensated for their payments into the pension system both in the form of transfers from workers and in the form of avoided climate damages. Subsequent cohorts of workers should also expect to receive less than they paid into the pension system. These decreases in the size of the transfers made from workers to retirees can continue until the point at which a cohort of workers secures a net benefit over its lifetime from any investments they make in mitigation capital. This is the point in time when there is no longer a problem of motivating these types of investments. In theory, there can be a stopping point for transferring the costs of mitigation to future generations despite the fact that the benefits of mitigation can be expected to extend very far into the future.[4] This looks like a clear method for taking on debt to finance investment in mitigation and effectively transferring the costs into the future.

## THE DIFFICULTIES OF COMPENSATING THE PRESENT

Changing the way we save for future generations only appears to be able to solve the problem of motivating the current generation in the straightforward way described above if current savings actually *aim* at passing on wealth to future generations. However, to a large extent the intergenerational savings effect of investments in the conventional capital stock appears to be a byproduct rather than the aim of these investments. Savers save to distribute their consumption over both the productive and unproductive years of their lives, to secure some of the gains of economic growth, and to pass on some wealth to their immediate descendants. Borrowers borrow to make productive investments that will bring them returns that are more valuable than the cost of borrowing. When the government borrows to invest in things like infrastructure, education, and health care, the time frames for returns are longer than for private investors. Still, if the government borrows to build a hospital, a university, or new roads the main aims are to use these goods now and to produce economic growth that will be beneficial in some way to taxpayers and their children.

The claim is not, of course, that the present's investments are in no way aimed towards the interests of future people. Those who engage in basic research may in part do so because it represents a good career for them.

Individuals and society may invest in such research because it is valued for its own sake. However, if there were no prospect of this research doing some good in the future we would surely invest much less. This is especially true for areas such as cancer research, but it also appears to be a general feature of our interests in the future (Scheffler 2013, 24). A real concern for the future must play some part in explaining why we sometimes invest in infrastructure designed to last for many generations. Taking resources away from things like cancer research or designing hundred year bridges and re-directing them towards mitigation may in fact be a more effective way of investing in the future. However, the large majority of investments in capital aim at benefiting the present even though they also often benefit the future as a by-product. Thus, asking the current generation to shift its investments in conventional capital towards mitigation capital is for the most part not a cost-free way for the present to produce better returns far off into the future. Rather, a motivationally challenged present needs to be compensated for not making the investments they currently make for more self-interested reasons.

If we go back to our pension scheme, it should now be clear that if current workers dedicate an extra 1 percent of their productivity to mitigation receiving an extra 1 percent of our children's productivity is not enough to compensate us in a sacrifice-free way. From the perspective of current workers and their children the value of their lifetime consumption bundles is greater in the BAU scenario compared to a scenario in which investments are shifted from conventional capital to mitigation capital. Resources are directed away from the types of consumption and investments that produce the best economic outcomes over the period that is relevant for current workers. When current workers become pensioners, they need to be compensated for the effect this decrease in the rate of growth will have on the size of transfers into the pension system compared to BAU. This means that our children will have to dedicate a larger percentage of their productivity into to the pay-as-you-go pension system than we did. This is only the first way in which reaching the no-sacrifice level is more difficult than it may at first seem.

Think of an economy that consists of a smoker and a room filled with asthmatics. The smoker internalises the benefits of smoking and externalises the costs. The asthmatics' lives are made almost unbearable by the suffering the second-hand smoke causes them. These social costs of smoking are much greater than the personal benefits the smoker enjoys. Let us suppose that the smoker is not moved by the plight of the asthmatics and that the only available option to eliminate the negative externality in this economy is for the asthmatics to compensate the smoker for quitting. He must be compensated to an extent that at least matches the benefits he enjoys from smoking. If the smoker quits smoking, the asthmatics will become amazingly 'welfare rich' compared to current conditions. However, this does not free up resources that can be used to compensate the smoker. The asthmatics become rich in the form of avoided asthma attacks. By assumption, the value of this form of wealth is extremely low from the perspective of the smoker. If the asthmatics are poor in other types of goods while the smoker has a very strong preference for smoking, a transfer to compensate the smoker for quitting will not be possible.

For a social planner trying to maximize welfare it is clear that permitting smoking in this economy is very inefficient. However, this is not how economists conceive of the way a negative externality can create inefficiency in a *market* that can be eliminated by a transfer that leaves no party worse off. Instead we have to see the value of smoking in terms of the smoker's willingness to be compensated for not smoking. In other words, the social benefit of smoking a cigarette is determined by how much we would have to pay the smoker so that he would be at least indifferent between the options of smoking the cigarette or taking the payment. The social cost of smoking is a function of the asthmatics willingness to pay to prevent smoking. When some agents' willingness to pay to avoid a negative externality is greater than the amount necessary to pay some other agents to refrain from creating this externality there is inefficiency in the market. A transfer from the negative externality takers to the externality producers generates a more efficient market outcome (Kelleher 2015, 71-73). In the smoking case I have described, we can assume that the asthmatics' willingness to pay is greater than the smoker's willingness to be compensated. However, the point to notice is that even if we have a large negative externality that is massively inefficient with respect to welfare outcomes it can also be true that there is no possible transfer between agents that could diminish the externality while each agent remains, at the very least, on their Pareto indifference curve. The extent to which an externality reducing transfer will be possible is dependent on the pollutees having access to goods that are candidates for transfer because they satisfy the polluters' willingness to be compensated.

Once we focus on the question of what the present appears to be willing to be compensated with to stop consuming GHG intensive goods it becomes clearer how large this alternative bundle of resources may have to be. What is at issue is the core of the current generation's consumption preferences and productive capacity. For example, effective climate mitigation may require moving largely to a vegetarian diet. The resources necessary to make such a transition are negative. Production of plant-based foods requires fewer resources than the production of meat. Small reductions in meat consumption are surely easy to compensate, but if we are aiming to compensate without having to wait for people to change their preferences (which is what the borrowing from the future proposal aims for) at some point the marginal willingness to be compensated for not being able to eat meat will become very low. Likewise, it may only take an annual investment of 1 percent of gross world product to mitigate climate change. However, getting to the no-sacrifice level may require a very large bundle of alternative resources to compensate the current generation for not being able to exploit emission-intensive goods that they have strong demonstrated preferences for.

The upshot is that in addition to compensating for differences in economic growth compared to BAU our children will also have to dedicate even more of their productive capacity to the pay-as-you-go pension scheme to make it worth the present's while to change the composition of its consumption bundles. Also note that the system calls for us to shift more of our consumption from our productive years to our non-productive years than we would normally choose to do. This is an opportunity cost and has to be compensated by more consumption in our non-productive years than we forwent in our productive years. This is a third additional cost that results in still greater shares of our children's productivity going into the pension system. Our children's children will in their turn have to dedicate even larger portions of their productivity to their parents than they did for us.

The point is not to suggest that it is more expensive than we think to mitigate climate change. Nor is the concern that there must necessarily be a very low elasticity of demand for GHG intensive goods that will make GHG emissions prices less effective than expected. The point is that even if emissions taxes effectively bring about desired shifts in consumption and investments it looks like it is more expensive than we think to compensate the current generation to the extent that they do not view these taxes as generating important sacrifices. It looks more difficult than expected to solve the problem of motivating agents to adopt effective carbon taxes in the first place. This is important because of limits to the amount of debt financing countries can engage in.

The typical story about limits to deficit spending is that the more debt a country has the more tax revenue they have to dedicate to servicing the debt, which raises the risk of default, which raises the interests rates at which governments can borrow, which in turn increases the revenue necessary to service the debt, which eventually makes further borrowing too costly. Given limits to how much debt governments can take on, whatever the mechanisms, it seems to follow that a generation that is unwilling to take on significant sacrifices to mitigate climate change is also going to have a strong preference for using debt financing for the sake of more present-oriented goods rather than more future-oriented goods. In other words, there is an opportunity cost here that looks like it is very difficult to compensate. Given that the present needs to be compensated for shifting its investment patterns, changing its consumption bundles, changing its consumption timing, and for the opportunity cost of dedicating scarce access to debt financing to future-oriented investments it should no longer be surprising that the current generation does not use debt financing to a substantial extent to invest in climate change mitigation. This is especially true given the magnitude of mitigation investments needed in comparison to existing deficit levels.

Net government deficits for the OECD countries in 2013 were 2.1 trillion US $.[5] In order to get onto a 2°C trajectory the International Energy Agency is calling for investment in addition to those needed simply to meet future energy demand of on average 1 trillion US $ per year to 2035 (International Energy Agency 2014, 44). Rogelj et al. (2013) estimate that an immediate global price on GHG emissions of US $40 $tCO_2e$, rising thereafter, would give us a 66 percent chance of keeping warming to 2°C. Global GHG emissions in 2011 were over 43,000 $MtCO_2e$.[6] This gives us over 1.7 trillion US $ in new costs. The co-benefits from such mitigation investments may be very large over the longer-term, while cost estimates may also be overly optimistic due to assumptions of full global cooperation and perfect policy implementation. There is obviously a lot of uncertainty about costs, but what is clear is that meeting these costs through debt financing involves extremely large shifts in how resources are being used compared to current patterns.

## EXPLOITING TYRANNY OVER THE FUTURE FOR THE GOOD

The combination of limits to the ability of governments to take on debt, incentives to use debt for present-oriented goods, preferences for GHG intensive goods over alternative packages of goods, and a system of investment in capital that largely aims at producing returns over the nearer term should make us question how big of a role borrowing from the future can play in addressing *motivational* obstacles to investing in mitigation capital. If this assessment is plausible we are back where we started. We see strong incentives for delaying mitigation investments and as a result it may be strategically important to identify ways of shifting the costs of mitigation to avoid dangerous levels of global warming.

In one sense, 'passing the buck' to the future has been a key strategy in climate politics for several decades. The 1997 Kyoto Protocol is regularly derided as having had far too weak commitments, covering far too little of global emissions, and as having been ineffective in bringing about emissions reductions that would not have occurred for other reasons. However, it is also widely understood that the protocol was weakly demanding in order to secure broad international participation and to make it possible to set up the institutional mechanisms for carbon trading and other flexibility mechanisms. The aim was to extend the regime in subsequent commitment periods with deeper reduction targets and more effective institutional mechanisms. This did not occur as envisioned, but the push within the UNFCCC process for more ambitious emissions commitments and expanded coverage continues. In the EU Emissions Trading System's (ETS) initial trading period between 2005 and 2007 the market was characterised by weak emissions targets, an over-allocation of free emissions allowances, and significant national flexibility in meeting targets (Parker 2011). Increased ambition and harmonisation of rules across EU member states, especially for the third trading period 2013–2020, has followed. However "temporary exemptions, compensations and procrastination of decisions" still create delays between decisions and the arrival of costs with these delays designed to help secure agreement (Muller & Slominski 2013, 1437). For example, the full shift in the third trading period to auctioning of emission credits will not be in place until 2027, while sectors deemed to be exposed to significant carbon leakage will have access to free allowances and the possibility of compensation.[7]

Creating temporal space between when policy makers adopt at decision and when the costs arrive, working with shorter-term flexibility and longer-term pre-commitments, and setting in motion path dependencies are all highlighted in the political science literature as important strategies for dealing with so-called 'super wicked' cooperation problems (e.g., Lazurus 2009; Ismer & Neuhoff 2009; Levin et al. 2012; Urpelainen 2012; Brunner et al. 2012; Jordan & Matt 2014). To the extent that these types of policies involve weak initial demands they do not respond to the urgency of the environmental threat. This is regularly and rightly criticised, but it should also be clear that such policies are often genuine strategic responses to real political obstacles. The borrowing from the future proposal aims to offer a better strategy than incrementalism by shifting costs to the future while at the same time eliminating procrastination. I have raised doubts that this approach is a powerful tool for mediating the politics of delay. The structure of the problem suggests that we need to identify ways to set in motion serious mitigation efforts but at the same time not require large changes in behaviour right now.

For example, those in political power now could commit the young to significant investments in mitigation. Designing these commitments so that they create a future financial liability for failures to make promised investments would exploit future decision makers' commitments to property rights regimes and the global financial system. Potential financial liabilities for failing to invest in mitigation would be a means to entrench mitigation commitments imposed by present governments onto future governments.[8] The only proposal along these lines I have been able to identify is the idea of governments issuing index-linked policy performance bonds where interest payments on the bonds are linked to GHG emission targets or financing targets for low-carbon energy. If the government fails to meet its mitigation targets there is a penalty in the form of higher interest rates to be paid to bond holders (Ekins et al. 2014, 168–70). However, governments have to already be committed to increasing the resources dedicated to cutting GHG emissions to pre-commit *themselves* in this way. What I am imagining is a policy that largely pre-commits future governments and thus places some temporal space between when the pre-commitment is made and when governments have to start making the mitigation investments. I have not been able to identify thinking in the economics literature that would specifically meet the criteria outlined above. As a result I am only able to briefly suggest a *type* of proposal that attempts to enforce commitments made today on future governments by creating a financial liability today that will materialise tomorrow if governments fail to mitigate.

The aim of the pre-commitment proposal suggested above is to reduce the level of bootstrapping involved in the more incrementalist approaches we currently have while at the same time taking seriously the possibility that governments will continue to be very wary about binding themselves to strong financial commitments over the short-term. The type of proposal under consideration is clearly flawed in that it does not respond quickly enough to the environmental threat. Thus, it should be understood chiefly as an insurance policy against the risk of an intergenerational pattern of perpetual delay.[9] Because I cannot provide a design for the strategy proposed above, this chapter is limited to assessing the normative case for this more explicit form of pre-committing the future. The purpose of such an assessment is to give some normative permission to think about new creative ways of pre-committing future publics that can better mediate the wickedness of time in the Anthropocene.

If our parents had committed us to financial liabilities for failing to invest in mitigation could we plausibly argue that we did not deserve this type of treatment? The more unjustifiable it appears to be for us to simply fail to mitigate climate change and the longer we delay serious action, the less plausible it is to question that it would have been justifiable for past generations to bind us to mitigation investments. If we deserve paternalistic treatment for our unwillingness and political incapacity to make meaningful investments in mitigation then so may our children. We have good reason to expect the next generation to do better than us from a moral perspective in various respects, but it is far from obvious that we should expect so much change that they will not also have very strong incentives to discount the interests of the future. Thus, the claim is not that the ways and extent to which people are motivated by moral considerations cannot change, but only that the incentives to discount the far future look particularly hard to change and that we should have some insurance against this problem.

When current publics try to pre-commit future publics the most common normative objection is that this is a form of political domination over the future by the present. This concern is usually raised against the constitutional entrenchment of some substantive public policy by the current majority with the aim of limiting the ability of future majorities to make public policy in this same area. If the present is able to democratically determine what the right substantive policy is without such obstacles why should the future be denied this same democratic power? Given reasonable disagreement about politics it is problematic for the current public to paternalistically safeguard future publics from following their own majoritarian will. On what grounds do current majorities think they have better access to answers about the policies that ought to be adopted than future majorities (Waldron, 1999, 255–82)? However the type of pre-commitment I am proposing is not an effort to protect the future against *itself*. Instead, pre-committing our children to mitigation investments is an effort to protect the *further* future from the *near* future. What we do is not to democratically adopt some measure for ourselves that we then think should be maintained in perpetuity. Rather we *fail* to adopt some measure that we think ought to be put into practice for the sake of future generations and instead pass on that commitment to the publics that will follow us. Surprisingly then, the combination of an unwillingness and inability to adopt just legislation with respect to the interests of future generations that we are currently witnessing appears to significantly improve the justifiability of present majorities paternalistically pre-committing future majorities.

The most serious objection to the idea that we should bind our children to mitigation costs is that it is a form of moral corruption. As Stephen Gardiner puts it,

> If the current generation favors buck-passing, but does not want to face up to what it is doing, it is likely to welcome any rationale that appears to justify its behaviour. Hence, it may be attracted to weak or deceptive arguments that appear to license buck-passing, and so give them less scrutiny than it ought.

It is the claim that the obstacles to political action are particularly severe in the case of climate change that makes buck-passing in a safer way seem reasonable. How this claim is deployed in our moral evaluations is what warrants more scrutiny.

Given the enormous amount of wealth and technological capability we currently enjoy it is not plausible to be sceptical about the prospects for action due to a sheer lack of capacity. Rather, it is a lack of the right kinds of motivations that prevents us from bringing the climate threat under effective political control. Binding our children to the costs of mitigation is presented as a way for us to live up to our obligations to protect the interests of future generations, albeit a very imperfect response. However, to make this move the present's moral failure to act is actually conceived of as an external condition that existing agents must take into account as we decide how to protect the interests of the future. It looks like I have perverted our blatant discounting of future interests into a moral justification for passing on the costs of mitigation! What can be said in response to this charge?

It is clearly moral suspicious to appeal to current political obstacles to justify cost shifting to the future. There is an incentive to exaggerate the obstacles one is complicit in creating because this provides moral cover for doing little now to address the problem. At the same time, it is also problematic to conceive of the present as a singular agent that can simply decide not to exploit some other agent, the future. Prohibitions on exploiting other agents are a basic feature of our normative theories and social institutions, and the message is that we need simply not to do what we normally expect agents not to do to each other. Yet, a generation is not a singular agent or any agent at all. Instead, what is required is to coordinate the actions of individuals, companies, communities, and governments all over the world. What we need to coordinate around is not some ubiquitous feature of common sense morality but something new. Agents must let the effects of their present actions on conditions far into the future outweigh their interests in securing goods here and now. We may have always depended on the idea of future generations to give meaning to our projects (Scheffler 2013), but we have not had to face the prospect of stark conflicts between many of our own unextraordinary projects and welfare in the distant future. If we look at the conditions in which agents have tended to be successful in protecting common pool resources it is clear that in the case of climate change these conditions are not satisfied (see Dietz et al. 2003). Climate change is the most difficult cooperative challenge humanity has faced to date and as a result it is reasonable to think in terms of second and third best options without being accused of blatant moral corruption.

In response to long-term threats like climate change political theorists often argue for institutional reforms that will eliminate the tyranny of the present over the future. The most common proposals are to have the interests of the future represented in some way in the democratic process or to constitutionally entrench respect for the interests of future generations. Instead of trying to address symptoms of the tyranny of the present over the future, it would be better, it is argued, to address the institutional sources of this injustice. However, it remains highly uncertain if institutional reforms of this type will go far enough fast enough to bring about effective mitigation policies. The argument of this section is that because we may have a limited window of opportunity to deal with the problem of weak incentives to invest in mitigation, we should also consider strategies that attempt to exploit the present's tyranny over the future for the good.

The argument above has defended the paternalism of binding the future to mitigation costs. However, once a case for paternalism is made it is appropriate to ask if other forms of paternalism are preferable. For example, one could imagine more or less paternalistic government policies that attempt to change present people's preferences so that they are more in line with the interests of future generations. The argument advanced here clearly does not demonstrate which policy responses are, all things considered, the best ones. Much depends on how pessimistic we think we should be about current political conditions. The type of proposal I have advanced is thought of as an insurance strategy against the risk of perpetual political inertia.

## CONCLUSION

There is already some debt financing of mitigation investments and debt financing would surely be a large part of extensive government efforts to mitigate climate change. There is also a good strategic and normative case for passing on the costs of mitigation investments to the future. If the present has strong incentives to pass on the costs of climate change to the future we should at least try to identify ways of passing on those costs in ways that best protect the interests of future generations. However, I have argued that the option to debt finance mitigation does not really resolve the basic problem of motivating agents to change their consumption and investment behaviours. This raises the question of whether or not there are other ways to pass on the costs of climate change in a 'safer' way. Strategic buck-passing and efforts to pre-commit future publics to increasingly demanding mitigation efforts are also already a key part of climate governance. I have suggested that we should consider more explicit pre-commitment strategies that bind the young today to large investments in mitigation over their productive lifetimes. If we are failing to overcome the tyranny of the present over the future then we should consider how we might exploit that tyranny for the good. My argument is not a moral endorsement of the present's domination over the future and it is not in conflict with the typical institutional reforms political theorists advance to reduce the present's discounting of future interests. Yet, given the severity of the political challenges we currently face and the severity of the consequences of global warming we should also be open to the possibility that we may need stronger measures to prevent a scenario in which we perpetually put off investing in climate security for the future.

## REFERENCES

Anadon, L. D., and J. P. H. Holdren. 2009. 'Policy for Energy Technology Innovation.' In *Acting in Time on Energy Policy*. Washington, DC: Brookings Institution Press.

Andreou, Chrisoula. 2007. 'Environmental Preservation and Second-Order Procrastination'. *Philosophy & Public Affairs* 35.3: 233–48.

Asheim, G. B. 2013. 'Competitive Intergenerational Altruism'. Accessed March 2014 at: http://folk.uio.no/gasheim/altrui03.pdf.

Broome, J. 2012. *Climate Matters: Ethics in a Warming World.* New York: W.W. Norton & Company.

Brunner, S., C. Flachsland, and R. Marschinski. 2012. Credible Commitment in Carbon Policy. *Climate Policy*, *12*(2) 255–71.

Dietz, T., E. Ostrom, and P. C. Stern. 2003. The Struggle to Govern the Commons. *Science*, *302*(5652), 1907–12.
Ekins, P., W. McDowall, and D. Zenghelis. 2014. *Greening the Recovery: The Report of the UCL Green Economy Policy Commission.* London: University College London. Accessed February 5 2015 at: http://www.ucl.ac.uk/public-policy/policy_commissions/GEPC/GEP-Creport.
Foley, D. 2009. 'The Economic Fundamentals of Global Warming' in J. M. Harris and N. R. Goodwin (eds.) *Twenty-First Century Macroeconomics: Responding to the Climate Challenge.* Northampton, MA: Edward Elgar Publishing.
Gardiner, S. M. 2001. 'The Real Tragedy of the Commons'. *Philosophy and Public Affairs,* 30, 387–416.
———. 2011. *A Perfect Moral Storm.* Oxford: Oxford University Press.
Hansen, J., M. Sato, P. Kharecha, and K. V. Schuckmann. 2011. 'Earth's Energy Imbalance and Implications'. *Atmospheric Chemistry and Physics*, 11(24), 13421–49.
Heath, J. 2013. The Structure of Intergenerational Cooperation. *Philosophy & Public Affairs*, 41(1), 31–66.
International Energy Agency 2014. *World Energy Investment Outlook.* Paris: OECD.
Ismer, R., and K. Neuhoff. 2009. Commitments through Financial Options: An Alternative for Delivering Climate Change Obligations. *Climate Policy*, *9*(1), 9–21.
Jordan, A., and E. Matt. 2014. Designing Policies that Intentionally Stick: Policy Feedback in a Changing Climate. *Policy Sciences*, *47*(3) 227–47.
Kelleher, J. P. 2015. 'Is There a Sacrifice-Free Solution to Climate Change?'. *Ethics, Policy & Environment*, 18(1), 68-78.
Lazarus, R. J. 2008. 'Super Wicked Problems and Climate Change: Restraining the Present to Liberate the Future'. *Cornell L. Rev.*, *94*, 1153.
Levin, K., et al. 2012. 'Overcoming the Tragedy of Super Wicked Problems: Constraining Our Future Selves to Ameliorate Global Climate Change'. *Policy Sciences*, 45, 123–52.
Luderer, G., V. Bosetti, M. Jakob, M. Leimbach, J. C. Steckel, H. Waisman, and O. Edenhofer. 2012. 'The Economics of Decarbonizing the Energy System: Results and Insights from the RECIPE Model Intercomparison'. *Climatic Change*, 1, 9–37.
Müller, P., and P. Slominski. 2013. 'Agree Now–Pay Later: Escaping the Joint Decision Trap in the Evolution of the EU Emission Trading System'. *Journal of European Public Policy* *20*(10), 1425–42.
Ostrom, E. 1990. *Governing the Commons: The Evolution of Institutions for Collective Action.* Cambridge: Cambridge University Press.
Parker, L. 2011. 'Climate Change and the EU Emissions Trading Scheme (ETS): Looking to 2020'. CRS Report for Congress, Congressional Research Service.
Rendall, M. 2011. 'Climate Change and the Threat of Disaster: The Moral Case for Taking Out Insurance at Our Grandchildren's Expense'. *Political Studies,* 59, 884–89.
Rezai, A., D. K. Foley, and L. Taylor. 2012. 'Global Warming and Economic Externalities', *Economic Theory*, 49(2), 329–51.
Rogelj, J., D. L. McCollum, A. Reisinger, M. Meinshausen, and K. Riahi. 2013. 'Probabilistic Cost Estimates for Climate Change Mitigation'. *Nature*, 493(7430), 79–83.
Rozenberg, J., S. Hallegatte, B. Perrissin-Fabert, and J. C. Hourcade. 2013. 'Funding Low-Carbon Investments in the Absence of a Carbon Tax'. *Climate Policy*, 13(1), 134–41.
Scheffler, S. 2013. *Death and the Afterlife*. Oxford: Oxford University Press.
Shue, H. 2005. 'Responsibility to Future Generations and the Technological Transition'. *Perspectives on Climate Change,* 5.
———. 2010. 'Deadly Delays, Saving Opportunities: Creating a More Dangerous World?' In S. M. Gardiner, S. Caney, D. Jamieson, and H. Shue (eds.), *Climate Ethics: Essential Readings.* Oxford: Oxford University Press.
Urpelainen, J. 2013. A Model of Dynamic Climate Governance: Dream Big, Win Small. *International Environmental Agreements: Politics, Law and Economics*,*13*(2), 107–125.
Vaughan, N. E., T. M. Lenton, and J. G. Shepherd. 2009. Climate Change Mitigation: Trade-offs between Delay and Strength of Action Required. *Climatic Change*, *96*(1-2) 29–43.

# NOTES

1. I would like to thank Catriona McKinnon for in-depth comments that helped me greatly improve this chapter. Earlier versions of this chapter were presented during 2014 at the ECPR Joint Sessions, the Nordic Political Science Association Conference, The Academy of Finland's Centre of Excellence in the Philosophy of the Social Sciences, and at the Global and Regional Governance and Political Theory research seminars at the Department of Political Science at Stockholm University. I would like to thank participants at these events for their comments, in particular Matthew Rendall, Dominic Roser, Blake Francis, John Broome, Robert Huseby, Jonas Tallberg, Magnus Reitberg, Säde Hormio, Kian Mintz-Woo, Simo Kyllönen, Jonathan Kuyper, Ludvig Beckman, and Göran Duus-Otterström. I would also like to thank Alan Mehlenbacher, David von Below, and Nick Rowe for answering some basic questions about the notion of borrowing from the future.

2. I remain agnostic on the question of what a fair distribution of mitigation costs between generations would be.

3. This conclusion appears to be true even where agents are narrowly altruistic in the sense of having strong preferences for securing high consumption levels for their children (Asheim 2013).

4. There is a question of whether or not the one could plan to reduce pay-outs to retirees without undermining the pension scheme. Workers facing the prospect that they will put more into the pension system that they get out cannot be excluded from the avoided climate damages that are supposed to make up for this difference. As such they have an incentive to decrease their inputs into the system to what they can expect to get out of it. This in turn gives cohorts prior to them incentives to pre-emptively decrease their inputs into the system. This dynamic could undermine the credibility of the scheme. Perpetually rolling over the debt could be a better way to ensure the credibility of the scheme. Normatively assessing such a strategy would be dependent on a theory of distributive justice between generations.

5. By country GDP figures were taken from OECD (2014), "Gross domestic product in US dollars", *Economics: Key Tables from OECD*, No. 5. DOI: 10.1787/gdp-cusd-table-2014-5-en. By country deficit figures were taken from OECD (2014), "Government deficit / surplus as a percentage of GDP", *Economics: Key Tables from OECD*, No. 20.DOI: 10.1787/gov-dfct-table-2014-1-en. The calculation excludes Chile, Mexico and Turkey.

6. WRI, CAIT 2.0. 2014. Climate Analysis Indicators Tool: WRI's Climate Data Explorer. Washington, DC: World Resources Institute. Available at: http://cait2.wri.org. Accessed October 1 2014.

7. See http://ec.europa.eu/clima/policies/ets/cap/auctioning/index_en.htm.

8. Ideally, the future holders of the corresponding financial entitlements would be those in poorer countries most vulnerable to the effects of climate change. These entitlements could thus serve as some level of compensation for failures to mitigate. However, one would also want agents holding rights to payment for 'failure to perform' to be in a strong position to defend these entitlements.

9. Rendall (2011) also argues that borrowing from the future should be seen as an insurance policy against a pattern of political inertia.

*Chapter Six*

# Informational Approaches to Climate Justice

Steve Vanderheiden

While scholars continue to debate the normative bases and precise demands of climate justice, the basic outlines of an adequate response to the threat posed by anthropogenic climate change are clear enough. Pathways to reducing global emissions such that average temperature increases are held to two degrees of warming this century require urgent and significant action, with windows closing on the opportunity to avoid more damaging impacts if societies do not begin the decarbonization process soon. Given the collective action dimension of international climate change mitigation efforts, whereby all are tempted to free ride in the absence of binding national emissions targets, an ambitious treaty framework backed by a regulatory regime capable of enforcing universal national emissions caps would facilitate cooperation in pursuit of this objective. However, such a treaty has thus far been elusive and does not appear to offer a mechanism for ensuring that mitigation results will be delivered in time to meet the demands of justice. A major question for climate governance and ethics is therefore *how* its decarbonization objectives might be achieved, rather than precisely *how much* these objectives require of various parties or *why* they are morally significant or required by justice.

Informational approaches to decarbonization may offer one such partial remedy to problems associated with climate change, since they promise to contribute toward carbon emission reduction goals through educational and reputational pressures, but are unlikely to comprise a fully adequate remedy to climate justice objectives on their own. My interest here is in how certain kinds of inadequate but perhaps marginally beneficial remedies might motivate actions that are justified on ethical grounds, make progress toward objectives that are likewise justified, and so be justified on the basis of their role in facilitating ethical actions or bringing about ethical ends.

Harnessing information to inform and mobilize ethical action requires the existence of a motive to act on principle or to avoid harming, where ignorance concerning the effects of one's actions can prevent this motive from appearing or guiding behaviour. Where ignorance can sometimes excuse actions that would otherwise be wrongful, provided that the ignorance is itself reasonable and not willful (Bell 2011), the role played by information is fairly straightforward: it can potentially cancel excusable ignorance for those exposed to it, informing agents of the potentially harmful effects of their actions or else leaving them to cause harm through culpable ignorance, and can as a result make agents more responsive to relevant facts concerning their conformity with the normative commitments. Suddenly receiving reliable information that shows a highly probable and causally direct link between some contemplated action and serious harm to an innocent victim would immediately cancel any excusable ignorance concerning that link, and so should mobilize strong moral reasons to avoid that action and avert the harm.

The connection between some actions and harm is not so direct, narrowing the culpability gap between the ignorant and the informed agent. Information about embedded carbon within goods and services that are permissibly consumed involves this indirect link between actions and harming, since the marginal differences in individual carbon emissions that would result from any single consumer becoming highly vigilant and reducing their personal carbon footprints as much as possible would only indirectly contribute toward a marginal increase in anthropogenic drivers of climate change, with no one's personal decarbonization efforts able to demonstrably avert any identifiable climate-related harm. However, Hiller (2011) argues from the marginal consequences of single polluting acts that avoidable emissions are *prima facie* wrong, despite the absence of palpable effects of single actions, and Nolt (2011) connects lifetime personal carbon emissions to morally significant harm, suggesting that a consequentialist basis for personal carbon abatement duties may be available. Other bases for personal decarbonization imperative, such as those involving 'mimicking duties' through which one reduces personal emissions to what would be required under a fair cooperative scheme when one cannot be brought about (Cripps 2013), those concerned with fair distributions of carbon emissions (Hayward 2006; Dobson 2006), or those seeking to combine personal emissions reductions with carbon offsets to achieve carbon neutrality (Broome 2012), can be more effectively pursued through informational efforts like product carbon footprint (PCF) labels, which allow agents to more effectively act upon such imperatives, and may lead some to consider adopting them as the result of the information they provide.

## ON INFORMATIONAL GOVERNANCE

Reliable information concerning the extrinsic qualities of consumer products, along with reporting requirements for large-scale polluters and resource users, are often viewed as necessary conditions for ensuring accountability with sustainability imperatives (Stephan 2003; Auld & Gulbrandsen 2010). The gathering and dissemination of information concerning releases of harmful pollutants and production of waste makes possible conventional 'command and control' anti-pollution regulation, and data tracking the use by various parties of scarce environmental goods or services like water, energy, or greenhouse gas emissions absorptive capacity enables their more sustainable and equitable allocation (Ramkumar & Petkova 2007; Vanderheiden 2009), though in neither case does pertinent information and transparency provide sufficient conditions for realizing these objectives. Regulatory approaches to pollution control rely upon such information to track compliance with permitted emissions, but require monitoring and enforcement of compliance in order to be effective. Information in such cases provides an external assist to the primary regulatory tool, but depends upon external standards for its implicit normative critique, and does little to motivate individual or collective behavioural change on its own.

The reporting and use of information for monitoring and enforcement of pollution control measures therefore differs from what Mol (2008) terms 'informational governance', which relies upon incentives internal to the environmental information and transparency system, whether in terms of allowing an agent to more effectively advance their own ethical commitments or by mobilizing reputational accountability against bad actors by publicly shaming them for allowable but dubious actions. A core premise of this approach is thus that disclosure and transparency requirements have independent effects upon behaviour, unassisted by regulatory incentives. The 'information turn' in environmental politics suggests some transformative potential of information and transparency in the absence of coercive regulations, with information providing feedback to persons or firms about their impact upon the world that is then mediated through norms and affects behaviour directly, rather than being dependent for its behavioural effect upon externally imposed standards or policy-based enforcement mechanisms. In broad terms, this is the focus of discussion in this chapter.

Why might anyone think that information gathering and dissemination programs, on the basis of their own processes or incentives and without disclosure revealing noncompliance with external standards or otherwise triggering conventional enforcement mechanisms, could affect significant change in the environmental performance of individuals, firms, or polities? Several explanations appear within environmental policy literature. Disclosure and transparency efforts have been identified as mechanisms for ensuring accountability among state and corporate actors (Keohane 2006), linked to broader trends away from secrecy in international politics (Florini 1998; Mitchell 1998), and applied to education-based efforts to improve civic competence (Mitchell 2011). Bartlett (1986) argues that the process of conducting a review and preparing and presenting an environmental impact statement (EIS), as is mandated under the U.S. National Environmental Policy Act (NEPA), embeds *ecological rationality* (Dryzek 1983) within state decision making processes, emphasizing the benefits of procedural commitment to information-gathering over the public pressure afforded by avenues of legal appeal that EIS mandates also offer. According to Bartlett, 'federal agencies were required by NEPA to improve, coordinate, consider, and recognize commitments, relationships, and environmental effects', which in effect required that they 'begin using procedural ecological reasoning in their planning and decisionmaking' (107). Elsewhere referred to as *reflexive* regulation (Orts 1995) and viewed as an aspect of reflexive modernization (Beck, Giddens & Lash 2003), this form of rationality is seen as better accounting for the ecological constraints upon and effects of state action. While Bartlett focuses upon the internal dynamics of information gathering and reporting requirements, as agencies are required to take into account additional impacts of their decisions and so recognize new values in the calculus by which those decisions are reached, others (Boström and Klintman 2008; Doran 2009) have focused upon how information and transparency requirements affect individual and firm behaviour as the information is mediated by other actors.

Because the public dissemination of information allows outside parties to hold polluters or resource users accountable for their environmental performance beyond what the law requires, Mol (2010, 135) suggests that 'transparency relates directly to power as it aims to democratize information and empower the powerless by providing them with one of the most powerful resources in current times: access to and control over information and knowledge'. This thesis concerning the empowering effects of information supposes that members of the public may be more likely and better able to challenge polluters either directly through consumer boycotts or other shaming actions (Stephan 2003), or indirectly by pressuring state regulators to enact stricter pollution controls (Cohen & Santakumar 2007). Beyond the potential empowerment of external stakeholders to hold polluters to account, Orts (1995) suggests that transparency can create incentive structures favorable to environmental performance-driven innovation through which firms can derive reputational benefits.

## PUBLICIZING POLLUTION DATA: EPA'S TOXICS RELEASE INVENTORY

Perhaps the informational program most lauded for empowering affected members of the public and creating incentives for industry to improve its environmental performance is the U.S. Environmental Protection Agency's (EPA) Toxics Release Inventory (TRI), which places online a searchable database of toxic chemicals released by industry and federal agencies, including mapping functions that allow users to view pollution sources by geographic area (Harrison & Antweiler 2003). Fung and O'Rourke (2000, 123) commend this pollution disclosure program for 'the ease with which a variety of users—ordinary citizens, public interest groups, state agencies, journalists, and those in industry—can use its data to quickly and easily rank industrial facilities along a rough dimension of environmental performance', thus affecting the share value of publicly traded firms and incentivising managers to do (or appear to do) better (Sabel, Fung, & Karkkainen 1999: 6). As a community 'right to know' provision, the TRI's online database of data concerning local releases of toxins is thought to empower stakeholders in addition to informing them, and to create an incentive structure through which performance beyond that mandated by existing state regulation confers additional reputational benefits. Presumably, this empowerment mobilizes existing concerns for personal safety on behalf of meaningful exercises in public control over sources or repositories of pollution, clarifying if not creating environmental values.

Fung, who co-directs the Transparency Policy Project, has more recently backed away from this more optimistic assessment of TRI's potential for public empowerment, but remains convinced of the program's potential benefits. Writing later with Weil, Fung, Graham, and Fagotto (2006, 171), he notes that some firms 'sought to reduce their emissions by engaging in pollution prevention strategies while others substituted chemicals or changes accounting practices in ways that improved reports without necessarily improving public health'. Although not discounting its empowerment and disciplining potential altogether, the authors here place TRI in a middle category of disclosure programs, which are 'insufficient to generate effective policy outcomes but can be made to work in tandem with other government actions to embed information in action cycles that produce congruent behaviours by disclosers' (175). Existing evidence on market responses to TRI data, they note, do not show that the system's reporting requirements have had significant effects on local residential patterns or community action, suggesting that members of the public 'do not consider toxic releases when they decide what neighborhood to live in, where to send their children to school, where to work, or in what company to buy stock', and thus that TRI's effectiveness 'has been more limited than it appears' (171). Nonetheless, they found that some firms were led to take proactive pollution-control measures in order to protect their reputations and avoid anticipated regulatory threats, with federal regulators increasingly responsive to the new information.

Work by Dingwerth and Eichinger (2010) also finds that the links between environmental disclosure and empowerment are often overstated. In a study of the Amsterdam-based Global Reporting Initiative (GRI), which is 'regarded as the world's leading voluntary scheme for corporate non-financial reporting' (76), they find little evidence that GRI's transparency efforts lead to greater civil society empowerment. While such policies 'may work where information needs are limited' and 'where the comprehensibility and comparability of reported information is not a major problem', they 'are unlikely to work in the same way where information needs encompass a whole bundle of indicators, where the quality of data requires a higher degree of 'literacy' on the side of report readers, and where issues of comparability are more complex' (91). Moreover, since those bad environmental actors threatened by disclosure and transparency programs that threaten to bring them negative public attention are most powerful where strong civil society groups that might potentially serve as a counterweight to them are absent, in such settings 'the corporate sector can "tame" transparency policies, reduce their transformative threat, and tailor the instrument to their own needs' (92). Transparency systems, that is, work best where civil society groups are already strong, which is also where they are least needed, while such systems can be readily coopted where civil society groups capable of holding bad corporate actors accountable are weak, rendering such systems least effective where they are most needed. In effect, the authors find that transparency systems empower the already-empowered, but fail to empower publics and potentially allowing polluters to hijack those systems where state regulatory capacity is also weakest, and vulnerability to environmental hazards the highest.

Aside from the paucity of evidence that online inventories of environmental hazards do in fact empower citizens in the way that advocates often claim, the increased access to information can have downside consequences in terms of the reactions that it induces, at least with regard to one kind of disclosure and transparency program (Langley 2001). Informational approaches like the TRI stress exposure risks, disseminating data about local environmental hazards, and so convey the dual message that one is vulnerable to harm from local sources of pollution but also potentially more empowered to minimize that vulnerability by virtue of knowing about it. Critics have questioned these claimed empowerment effects, however. Etzioni (2010) argues that environmental regulations have an 'expressive function' in declaring community norms against important hazards by controlling their causes, whereas non-binding transparency rules imply that the threat in question is 'less consequential than if the activities or products at issue are banned or their provision is required' (15). Similarly, Szasz notes that information about environmental risks like that disseminated through TRI generate a potentially disempowering and depoliticizing reception in many (Szasz 2007, 2–4). If the environmental impacts that persons are informed about concerns risks to which they may be exposed by virtue of some of their choices, such as where to live and work, their reaction may be to adopt a defensive posture with regard to other choices that they might more readily alter, such as what to eat, drink, or wear. As Szasz notes, this defensive reaction is apolitical and not very constructive, but it also reinforces an inward-focused orientation in which environmental information erodes the normative commitment to sustainability upon which the most promising informational approaches depend.

## ECO-LABELS AS MARKERS OF EXTRINSIC PERFORMANCE

But there is another kind of information that at least in principle might be able to yield the sort of socially-oriented concern for sustainability that is needed for such approaches to rival regulatory ones in their effects. The reaction that Szasz describes might follow from fear of the intrinsic effects of certain consumer goods, like 'pink slime' in ground beef or bovine growth hormone in dairy products, prompting consumers hearing about such additives to seek out 'natural' or other putatively safer alternatives, or from general knowledge about other nearby sources of contamination, provoking this defensive reaction that manifests in actions over which persons have some control. However, one would expect a quite different reaction to information about more widely distributed extrinsic effects that result from the manufacture, use, or disposal of the products we consume—about our global rather than very localized environmental impacts. Information of this second kind draws attention towards social rather than personal risks from certain kinds of products or activities, cast in terms of social or environmental costs to one's polity or the larger world, while identifying negligible personal impacts and offering no reason to modify one's behaviour from strictly selfish motives.

This kind of information can be conveyed through eco-labels, which focus on social and environmental impacts and convey information about the extrinsic effects of commodity choices, most of which have no discernible impact on the consumer purchasing them other than the kinds of reputational or status benefits that such consumption entails, thereby appealing to other-regarding concerns rather than consumer self-interest (Kaiser & Edwards-Jones 2006; Boström & Klintman 2008). Buying certified Fair Trade or USDA Organic coffee rather than uncertified alternatives promises no personal benefit to the consumer, either in terms of better taste or lower personal risks associated with consuming the product. Rather, it promises better working conditions for growers and pickers, and better prices paid to both, along with (with organic certification) reduction in local impacts from sludge or synthetic chemicals used as fertilizers or pesticides. Given the altruistic nature of the impacts that they highlight, eco-labels appeal to those for whom such impersonal effects are important. In other words, the value of eco-labels to consumers depends upon the prior existence of environmental values or concerns for social justice beyond that which can be grounded in self-interest. Preferences for credence goods may not originate from information about 'sustainable' or 'ethical' options alone—informational approaches are primarily viewed mobilizing rather than constructing the values on which they depend—but may be activated, applied, and strengthened by such efforts.

To be effective, informational efforts that rely upon product or firm certification or which report data that can be used to compare the relative impacts among alternatives must be credible to the consumer as well as provide pertinent and usable information that facilitate informed consumer choices (Auger et al. 2003). An instructive example is the food label, which can contain information not only about intrinsic properties like ingredients and nutritional value (which involves a distinct informational dynamic), but also extrinsic effects of its production on the larger world. Labels certifying primarily extrinsic properties like 'GMO-free' foods have generated more controversy than those reporting purely intrinsic properties like caloric content (Gupta 2010), as the result of food industry opposition to such labels and World Trade Organization standards that view such process-based labels as constituting an illegal trade barrier. Such controversy owes in part to their potential to empower consumers to use their purchasing power to oppose processes that may not qualitatively change their food (though this remains a controversy with GMO foods) but which can have palpable effects in the world.

Since a significant part of the market appeal of such goods lies in their claim to a more socially or environmentally benign supply chain impact, products certified and marketed in this way are known as *credence goods*. Third-party certification schemes promise consumers of credence goods that specified social or environmental standards have been met, providing an additional source of value beyond that intrinsic to the commodity itself. With Fair Trade coffee, for example, NGO certification requires that growers be paid a floor price above market rates for beans, and that production processes meet basic labor and environmental guidelines. Critics point out that the required floor price is still quite low, that most of the added value from fair trade coffee accrues not to growers but to the relatively affluent roasters and distributors of fair trade beans, and that requirements to form cooperatives may have hurt some growers (Philpott et al. 2007; Jaffee 2012). Since large corporations like Folger's and McDonald's have used their buying power to eliminate middlemen and so gain fair trade status for their coffee, critics worry about the effects on small suppliers of this dilution of fair trade status, and in response have proposed expanding the binary certification with a tiered system that recognizes varying levels of support for growers or commitments to sustainable processes. Nonetheless, Fair Trade certification represents a higher standard than non-certified alternatives, appealing to the 'ethical consumer' to treat this as a source of the product's value, alongside its intrinsic properties.

In a study of market demand for Fair Trade coffee, Margaret Levi and April Linton characterize 'green' or 'ethical' consumerism as maintaining that 'purchasing power is used to promote moral ends, goals that serve the material interests of others often at a cost (albeit sometimes relatively minor) to the consumer' (2003, 407), where the goal is to change behaviour 'by transforming individual tastes and preferences' and inculcating 'the norm that people in prosperous countries should factor global social justice into their buying decisions' (419). Their findings, however, cast doubt on the transformative potential of such an approach. They find that few consumers are willing to buy certified beans unless they also taste good, and then are only willing to pay a small premium for the credence good that certification represents. If the extrinsic social and environmental effects of Fair Trade certification efforts constitute the basis for stand-alone reasons for consuming one product rather than another, they suggest, the value of the credence good that certification provides is relatively small. Likewise, consumers willing to pay only a small premium for goods with purportedly smaller social or environmental impacts upstream in the supply chain may not be acting from ethical motives of harm avoidance so much as seeking to assuage guilt related to their consumption practices, calling into question whether such practices warrant the label of 'ethical' in the first place.

Informational efforts to inform consumers of the social and environmental impacts of their purchasing decisions rely upon a dynamic by which latent social and environmental values may be activated and directed toward more just or sustainable consumption behaviours, and perhaps instantiated and strengthened by presenting evidence of global problems with which existing consumption patterns are causally linked. Information that is gathered, processed, and disseminated through such efforts may have value beyond its use in product certifications and labels, consumer education campaigns, and transparency programs, including its use in lifecycle analysis programs that promote sustainability objectives by minimizing waste and in regulatory monitoring and compliance. While acknowledging its limits as an agent for widespread social and environmental change, I shall indirectly consider the transformative potential of information by focusing more narrowly upon a key dynamic of the workings of informational efforts, along with a normative question that this dynamic suggests; namely, whether persons or states have a responsibility to recognize available information about the broader impacts of consumer choices, and what follows for politics from an affirmative answer to this question.

Information has this transformative potential on a small scale insofar as some persons notice and cognitively assimilate it, revealing conflicts between what that information indicates and their considered value judgments, thereby promoting reflection on and revision of existing preferences and thus behaviour change as a result. Those, for example, who are opposed to animal cruelty but unaware of which products rely upon it might benefit by information systems that accurately and credibly make this distinction. They might personally be able to exercise greater moral integrity by avoiding consumer behaviours that conflict with their considered ethical judgments, thus becoming more ethical consumers, as the result of information and transparency efforts. Prospects for reducing the occurrence of animal cruelty might depend upon decreases in market demand for products from sectors in which cruelty is pervasive, rather than product differentiation through certification schemes within those sectors. The potential for bringing about change on a wider scale depends either upon the number of consumers affected by informational efforts in the manner described above, or through the transformation of public norms, which condition preference formation and so can have force by instantiating rather than merely mobilizing relevant values. Evidence of informational campaigns creating public distaste for certain product types rather than merely mobilizing latent preferences against those goods can be seen in the rapid public shift away from Canadian seal fur following Greenpeace anti-sealing campaigns (Dauvergne 2010), but evidence for impacts upon meat consumption patterns from campaigns against veal and other factory-farmed animal products has been elusive (Tobler, Visschers & Siegrist 2011).

## CERTIFICATION, TRANSPARENCY, AND RESPONSIBILITY

The critic of informational approaches is surely right in suggesting that serious global problems like climate change cannot be effectively addressed through such voluntary measures alone, but the defender of the informational approaches on which green consumerism depends can concede this point without abandoning the transparency project altogether. So long as it is treated as merely one mechanism among several for encouraging persons to behave more responsibly in the context of climate change or other social or environmental consequences of current consumption patterns, opposition to regulatory measures is not a necessary feature of the defense of voluntary ones. Informational approaches promise to develop a resource that is just as useful for holding persons responsible through regulatory efforts as it is for inducing them to voluntarily take responsibility for their actions. If states want to impose carbon budgets on their citizens or attach a price to carbon through taxes or trading systems, information about the carbon emissions resulting from various products or activities is needed in order to monitor and ensure compliance with regulatory efforts, as are the lifecycle analyses through which such information can be gleaned. If persons are to comply with self-imposed personal carbon allowances or satisfy voluntary carbon neutrality objectives (Broome 2012), they likewise need information about their current footprints and the effects on them of alternative choices so that they can make informed carbon budgeting decisions. Without it, they cannot take responsibility or be held responsible for their carbon emissions (Vanderheiden 2011), since their carbon footprints cannot be compared against their carbon budgets. If they are to develop the virtues of ecological citizenship associated with seeking to claim only equitable shares of ecological space (Dobson 2006), they require basic information about their personal consumption impacts along with comparative data on per capita footprints elsewhere and system-wide ecological capacities. Such information also risks disempowering persons when ethical demands to sharply reduce their footprints appears overly demanding, as with data showing U.S. per capita carbon footprints to be well in excess of what is required to meet two degree maximum temperature increase goals, so both positive and negative effects upon mitigation ambitions must be kept in mind as relevant data is collected and presented for public consumption.

What sort of information is required, and how does it work to facilitate these mechanisms of political accountability and personal responsibility? An instructive example is the food label, which can contain information about its various intrinsic or extrinsic attributes. Food labels now reveal information about a product's intrinsic nutritional properties: its ingredients, calorie and fat contents, percentage of a day's recommended dose of vitamins and minerals contained in each serving, and so on. In some cases, warnings are issued on labels, as with alcoholic beverages and pregnancy. Food labels are useful for those following strict dietary guidelines, like prohibitions upon animal products, as well as for those seeking to limit but not entirely avoid things like calories or carbohydrates. Labels that certify some combinations of intrinsic and extrinsic attributes like those identifying kosher or organic foods are both highly useful to those for whom such attributes are important and relatively uncontroversial. On the other hand, labels certifying primarily extrinsic properties like 'GMO-free' foods have generated more controversy (Gupta 2010), as the result of industry opposition to such labels and free trade standards that view such process-based labels as constituting an illegal trade barrier. Controversies over labelling schemes owe in part to their potential to empower consumers to use their purchasing power to oppose processes that may not qualitatively change their food (though this remains a controversy with GMO foods) but which can have palpable effects in the world.

Binary certification schemes like Fair Trade and the anti-sweatshop No Sweat label make no distinctions among products earning the label and provide no information about products that lack certification. Graduated schemes offer more detail, as in the USDA Organic label's three levels of certification for products made with 70 percent, 95 percent, and 100 percent organic ingredients, but convey no information about uncertified products. More universal and linear labelling schemes could potentially provide far more useful information, since they would allow more meaningful comparisons to be made within categories of goods or services as well as between such categories. Persons using them to advance their ethical commitments could not only more accurately determine how best to avoid harming within the constraints of their current consumption preferences, if adopting carbon neutrality pledges or engaging it 'mimicking duties' that require cuts personal carbon footprints proportional to those required for collectives under climate justice imperatives (Cripps 2013), but could also evaluate those preferences on the basis of their relative impact compared to alternatives. Energy efficiency labels, for example, enable easy comparison among alternatives within some product category in terms of energy efficiency, which concerns their intrinsic properties in affecting energy costs to the consumer as well as extrinsic properties in assessing relative environmental impacts. Product carbon footprint (PCF) labels, which report only upon the extrinsic property of a product's carbon footprint, can similarly generate incentives for manufacturers to decarbonize their product lines while also empowering consumers to act on their environmental values, and perhaps also to strengthen such values through a process of reflective equilibrium between one's lifestyle or consumption preferences and their environmental commitments.

PCF labels can influence consumer choices or inform institutional purchasing decisions by revealing comparative information about a given product's impact on climate change, but can potentially also be a valuable diagnostic tool in encouraging firms to conduct life-cycle analyses that identify efficiencies in their supply chains. As Vandenberg, Dietz, and Stern observe:

> Labelling also may induce firms to reduce their emissions in ways that lower their costs, enhance their reputations and make them more supportive of governmental policy measures that reinforce their emissions-reducing actions. This easily overlooked effect of carbon labelling will occur to the extent that firms respond to generalized concerns about brand reputation even if consumers only demonstrate limited willingness to pay for lower-carbon goods. Indeed, it seems that many firms have overlooked supply-chain efficiencies, and are not acting on substantial opportunities to cut costs and reduce emissions. Developing the data to underpin carbon labelling can identify and highlight these potential savings and spur changes in production and distribution

> throughout the supply chain; an effect that may be a more potent incentive than the immediate impacts of consumer choices. Industries have responded similarly in the past. (2011, 5)

As with other forms of life-cycle analysis through which supply chain impacts are estimated through comprehensive inventories of production-related material inputs and environmental releases, the procedural requirement entailed by such disclosure programs involve substantial initial costs. However, such costs can be partly justified by the economic benefits to the firm of potential for realizing efficiency gains in reducing materials or energy use or waste production, combined with the reputational and environmental benefits of successfully pursuing these. Notably, only some of these benefits depend upon consumers being willing to pay more for green products.

Beyond these conditioning effects upon firms, however, perhaps the most promising element of PCF systems concerns the effect upon norms of transparent environmental impacts. As Vandenbergh and Steinemann note of carbon neutrality pledges, which are voluntarily taken by persons or groups but facilitated by carbon footprint calculators, the goal of achieving carbon neutrality (which requires one's personal carbon emissions to be balanced by offsets) 'enables individuals to take personal responsibility for their contributions to climate change without reliance on uncertain or shifting estimates of the necessary reductions of others' behaviour' (2007, 1720). Apart from encouraging such voluntary efforts at decarbonization, along with allowing more robust measurement of progress toward carbon neutrality goals, Vandenbergh and Steinemann suggest that PCF systems also provide 'information that activates norms', which they suggest 'may be necessary for more traditional regulatory schemes to be politically viable' (1726). In fostering an ethos of what Dobson describes as 'thick cosmopolitanism', which he describes as 'identifying relationships of causal responsibility' that 'trigger stronger senses of obligation than higher-level ethical appeals can do' (2006, 182), PCF labels and the personal carbon accounting they encourage can not only promote greater ethical concern or reflection, improving individual behaviour outside of any coercive policy tools that incentivize sustainability, but it can potentially also generate the necessary public support for developing supplemental policy approaches. By starting to track information like product carbon footprints now, moreover, carbon leakage that results from emissions that are embodied in trade can be more readily identified, with national mitigation targets adjusted to reflect not where greenhouse pollution occurs through production processes but where demand for or consumption of carbon-embodied goods occurs (Davis, Peters & Caldiera 2011), leading to better state-level carbon accounting. Measures ensuring that no person or state can artificially shrink their carbon footprint merely by shifting the production of high carbon-embedded goods across borders help to reinforce equity norms, which depend upon agents being held responsible for the emissions they cause, while the sort of carbon leakage that results from the inability to track carbon through trade can undermine these norms by allowing some to shift the responsibility for their consumption impacts onto others.

## INFORMATION AND INTEGRITY

How do informational approaches promise to bring about preferences transformation and behavioural change, and what does this process imply for questions about responsibility? The moral psychology of informational approaches operates in those markets where the sale of less egregiously harmful products than some might otherwise consume is allowed, with agents facing an ethical dilemma when confronted by such information. Here, information is meant to appeal to the value judgments of some (but not all) persons in their capacities as consumers, and can report positive or negative attributes of products or the processes by which they are made and brought to market. Of primary importance to ethical decision making are those extrinsic impacts of goods or services upon other people or the environment, which persons *qua* consumers must assimilate and combine it with other information about the relative merits of alternative lifestyle and consumption preferences, forming and revising those preferences through reflection upon their considered value judgments. Informational efforts are thus linked to behaviour through personal values and social norms, and while behavioural change may be the primary policy objective from of such efforts, the focus here shall be upon the normative issues that surround the interaction of information with personal values and social norms.

A core component of this interaction is the psychological need for or moral commitment to integrity, which requires a kind of reconciliation between one's personal values and actions. Stephen Carter casts integrity as a 'pre-political' virtue 'without which other political views and values are useless' (1996, ix), suggesting its foundational role in normative political problems but also setting it apart from more substantive value commitments. Integrity doesn't presuppose any substantive social or political ends, but allows for their pursuit. Those exhibiting integrity in their personal or public lives avoid acting in ways that violate their deeply held views about right and wrong and seek a kind of reflective equilibrium between their value judgments and actions, revising both to bring them in accord with one another. As a psychological imperative, they are troubled by contradictions between their actions and beliefs, where they might not be true to themselves. As a social imperative, those with integrity are concerned with doing them as they outwardly represent themselves as believing and appearing to others as such. Insofar as the dissemination of information about the social and environmental effects of commodity options induces changes in consumer behaviour, the moral psychology at work involves integrity, in this inward private or outward public sense. Coming to know certain facts, whether that information was intentionally sought or inadvertently acquired, brings about this discomfort, with changes in one's consumer behaviour as one means of reconciling the conflict between values and action.

Other means for resolving such inner conflict are also available, and can work against the aims of informational efforts. The oft-observed phenomenon of youthful idealism giving way to hardened cynicism or self-oriented materialism could be explained by the gradual revision of social and environmental values rather than consumer preferences, where persons maintain their integrity by simply narrowing the scope of their concern or amending their moral commitments based on a revised assessment of what is possible, as the opportunity costs for maintaining that idealism with integrity increase. In addition, minimal cleverness is needed to rationalize one's existing consumer behaviour, convincing oneself that the gap between one's values and behaviour is not as large as it is in fact, often in ways that border on self-deception but nonetheless satisfy the psychological need to avoid cognitive dissonance. By adopting a self-serving skepticism about product information or the social and environmental problems with which it is linked, or by selective attention to data based on its propensity to validate existing behaviour, persons can maintain integrity while allowing wide gaps between their values and behaviour by convincing themselves and/or others that those gaps are in fact small or nonexistent. The tenacity with which we sometimes rationalize our consumer behaviour reveals the power that our integrity imperative exercises on us, along with illustrating one of our cognitive defenses against that imperative compromising our narrower self-interest. Insofar as information regarding the impacts of our actions can undermine this self-serving rationalization and so better hold our behaviour accountable to our values, transparency can increase the tension between our desire for integrity and our preference to avoid critically reflecting upon and potentially reforming our current consumption practices.

The internal dynamic in which integrity features is linked to issues of ignorance and responsibility, which raise normative questions about culpability for willful ignorance in the face of available information. If persons have some moral responsibility to seek out information concerning the social and environmental impacts of their actions and choices, to reflect on this information, and to revise the values and/or behaviour accordingly—where integrity plays the pivotal role in prompting our reflection and revision upon cognitive recognition of the information in question—then they cannot cope with the demands of integrity by adopting a self-serving but deceptive skepticism or maintaining a willful ignorance. A genuine moral commitment to integrity, as opposed to having a psychological need for resolution of inner conflict, could not be maintained disingenuously or through self-deception without being self-undermining. Persons could still revise their values in light of information that showed them to contradict those values in some actions that they have other reasons for seeking to maintain, so they need not necessarily revise consumer behaviour in the face of cognitively recognized information, but some popular strategies for coping with the gaps between values and actions that integrity seeks to close would be off limits, perhaps as inconsistent with integrity itself. A vegan might revise her outright ban upon animal products upon learning about what she regards as humane dairy practices, for example, but a self-serving but uninformed skepticism about the existence or perniciousness of sweatshop labor could not be adopted without violating integrity itself, even if such adoption allowed greater unfettered consumption opportunities. In this sense, information couples with integrity to guard against what Gardiner (2011) terms 'moral corruption', through which agents maintain willful ignorance in order to avoid incurring costs to themselves that accrue from doing as they ought.

This kind of responsibility, through which persons are made to view their actions in an interdependent global context, helps to define the notion of cosmopolitan citizenship (Dobson 2006), through which persons concern themselves with the global effects of their local actions, and has implications for informational governance to be explored below. As Young suggests, it may ask persons 'to reflect morally on the normal and hitherto acceptable market relationship in which they act' (2004, 378), challenging norms where necessary rather than seeking to hold specific agents responsible for harm caused by a combination of economic forces and overly permissive norms. Even where persons do not directly cause harm, which might result instead from the aggregation of many small contributions by many and multifarious acts, they can be viewed as complicit in harmful outcomes that they cannot prevent from occurring (Kutz 2000).

Several problems confound the transformative potential of information from having the transformational effects described above. Accurate information concerning the social and environmental effects of commodities whose production relies upon complex global supply chains can be hard to gather (Conca 2001), as these are typically deliberately shrouded in opacity and distant from those whose disapprobation would most threaten them, and greenwashing efforts can simultaneously frustrate data collection and undermine public confidence in its accuracy and objectivity. Absent credible and comprehensive information that can be gathered and consolidated, little benefit results from its effective dissemination. Supposing that this obstacle can be overcome and a credible and comprehensive dataset assembled, further problems with the presentation and dissemination of information arise, as overly complex data reporting can confuse or otherwise overwhelm its intended audience while oversimplified presentation can minimize its informational value and trivialize the range of impacts that some products cause by myopically focusing upon a single one. But supposing that this obstacle is likewise overcome, and good data is well and widely disseminated, a final problem is likely to face informational efforts: those exposed to this potentially transformative information may choose simply to ignore it, preferring to remain ignorant of the social and environmental effects of their consumer actions and behaviour, despite the low costs of rectifying this ignorance and acting accordingly.

Certainly, humans are generally disinclined to seek out information that might make them complicit in harmful outcomes that they cannot prevent from occurring. Many, to paraphrase Peirce, cling with tenacity to those beliefs that best fit with their views of the world and of themselves, often in the face of what should be reasonable suspicions about the accuracy of those beliefs. They rationalize their past actions and justify their future ones by stubbornly refusing to consider that those actions may contradict their core value commitments. These are, I think, basic facts about human nature: that many (though not all) persons are uncomfortable with the cognitive dissonance that results from conflicts between our normative and empirical beliefs. Regardless of how well-founded our normative beliefs are at the time that we form them, once formed we are reluctant to revisit them, which we may be forced to do if we admit contrary facts into our cognitive field. But tendencies latent in human nature cannot be self-justifying (Estlund 2011, 220–21), and we need further reasons than the desire to simultaneously avoid cognitive discomfort and the costs associated with being a good cosmopolitan citizen.

## CONCLUSION: INCREASING THE IMPACT OF INFORMATION

In contrast to disclosure programs like TRI that offer stakeholders information without allowing them the agency to grant their informed consent to the exposure risks that its data might reveal or empowering them to make choices that signal their approval or disapproval of high or low levels of environmental performance, the disclosure contained within an eco-label can potentially harness the agency of concerned stakeholders to make choices that support good environmental performance. All else being equal, eco-labelling has more transformative potential than TRI.

However, meaningful opportunities for agency are clearly not a sufficient condition for eco-labels to have the transformative potential described above, and so it might not be the case that all else is, in fact, equal. Most notably, eco-labels rely upon a different and less powerful set of motives for change, since their concern is not with disclosing information regarding local exposure risks, which residents may seek to avoid when evaluating their residential options or in deciding whether to move away from riskier locations, but is rather with the often distant effects of everyday choices that have little or no discernible impact upon those empowered to choose on the basis of the information they convey. Here, agency trades off against urgency, where persons are more empowered to take the kinds of actions that matter less to them. Concern for others, including the kinds of environmental and social impacts that could be conveyed through eco-labels, and perhaps inescapably less salient to most preference orderings than is self-interest, but this need not deter inquiry into the potential for harnessing the former in the service of defensible ends as well as the latter.

Whatever else they include, one way to increase the incentive effect of eco-labels is to take a lesson from the logic of online inventories. We might ask: why do we require polluters to publicly disclose their emission records, or manufacturers the environmental impacts of their production processes? As Gupta notes, the logic is the same in both cases, even though the former is a common and widely-accepted mandate while the latter is not. Disclosure programs, she writes, have three primary purposes:

> First, a normative right to know of recipients as an end in and of itself; second, it may seek to further various procedural ends, such as enhanced participation or choice of recipients, or enhanced accountability of disclosers; and finally, disclosure may seek to further substantive ends such as environmental improvements, sustainable resource use or risk reduction. (2010, 33–34)

Information about local environmental hazards and pollution sources, as is provided through programs like the TRI, can be seen as accomplishing all three of these: satisfying the 'right to know' demands of an affected public whose health is putatively being protected by pollution control agencies that through this protective responsibility must keep residents informed about any known risks; holding polluters accountable by disclosure combined with the empowered resistance to excessive risks that such disclosure enables; and creating incentives for improved performance by publicly shaming bad environmental actors and implicitly commending good ones. As Gupta suggests, the logic of transparency is the same for voluntary programs like eco-labels as it is for regulatory ones like online pollutant inventories of the TRI. Reliance upon purely voluntary eco-labels or certification programs only captures half of the reputational benefits noted above, since the voluntary nature of such systems entails that only potential beneficiaries of their reputational effects will opt in. Insofar as the public has the right to know about bad as well as good products and firms, and bad actors deserve to be held accountable or suffer reputational sanctions for their poor performance along with good actors benefitting by their better performance, labels or certification systems ought to be required of all products and firms, not merely the good ones. Eco-labels could wield the stick of bad publicity for bad performance along with the carrot of good publicity for its opposite, furthering the objectives of pollution reporting systems by allowing for pressure to be placed on polluters at the point of sale in addition to the end of pipe. By including linear rather than binary information and by doing this for a greater range of goods and services, as PCF and other carbon measurement and disclosure schemes promise, the utility of informational approaches in tracking conformity with either internally or externally imposed standards increases, as do opportunities for greater reflexivity and preference transformation in the way that persons interact with their environment. Information gathered and disseminated through such self-governing systems could also be of potential use to effective carbon accounting or pricing systems, which are able to supplement the educational and mobilizing value of voluntary programs with the binding force of law and policy, using much of the same information for different purposes. Together, the imperatives of climate justice that depend upon the recognition of and aspiration toward quantitative targets based in principles of distributive equity might more effectively be realized.

## REFERENCES


Auger, Pat, Paul Burke, Timothy M. Devinney, and Jordan J. Louviere. 2003. 'What Will Consumers Pay for Social Product Features?' *Journal of Business Ethics* 42: 281–304.

Auld, Graeme, and Lars H. Gulbrandsen. 2010. 'Transparency in Nonstate Certification: Consequences for Accountability and Legitimacy'. *Global Environmental Politics* 10: 97–119.

Beck, Ulrich. 1992. *Risk Society: Towards a New Modernity*. Thousand Oaks, CA: Sage.

Beck, Ulrich, Anthony Giddens, and Scott Lash. 2003. *Reflexive Modernization: Politics, Tradition, and Aesthetics in the Modern Social Order*. Palo Alto, CA: Stanford University Press.

Bell, Derek. 2011. 'Global Climate Justice, Historic Emissions, and Excusable Ignorance'. *The Monist* 94(3): 391–411.

Boström, Magnus, and Mikael Klintman. 2008. *Eco-Standards, Product Labeling, and Green Consumerism*. New York: Palgrave Macmillan.

Broome, John. 2012. *Climate Matters: Ethics in a Warming World*. New York: Norton.

Carter, Stephen L. 1996. *Integrity*. New York: HarperCollins.

Cohen, Mark A., and V. Santhakumar. 2007. 'Information Disclosure as Environmental Regulation: A Theoretical Analysis'. *Environmental and Resource Economics* 37: 599–620.

Conca, Ken. 2001. 'Consumption and Environment in a Global Economy'. *Global Environmental Politics* 1: 53–71.

Cripps, Elizabeth. 2013. *Climate Change and the Moral Agent: Individual Duties in an Interdependent World*. Oxford, UK: Oxford University Press.

Dauvergne, Peter. 2010. *The Shadows of Consumption: Consequences for the Global Environment*. Cambridge, MA: The MIT Press.

Davis, Glen, Glen P. Peters, and Ken Caldiera. 2011. 'The Supply Chain of $CO_2$ Emissions'. *Proceedings of the National Academy of Sciences of the United States of America* 108: 18554–59.

Dingwerth, Klaus, and Margot Eichinger. 2010. 'Tamed Transparency: How Information Disclosure under the Global Reporting Initiative Fails to Empower'. *Global Environmental Politics* 10: 74-96.

Dobson, Andrew. 2006. 'Thick Cosmopolitanism'. *Political Studies* 54: 165–84.

Doran, Caroline Josephine. 2009. 'The Role of Personal Values in Fair Trade Consumption'. *Journal of Business Ethics* 84: 549–63.

Dryzek, John S. 1983. 'Ecological Rationality'. *International Journal of Environmental Studies* 21: 5–10.

Estlund, David. 2011. 'Human Nature and the Limits (if any) of Political Philosophy', *Philosophy & Public Affairs* 39: 207–37.

Etzioni, Amitai. 2010. 'Is Transparency the Best Disinfectant?' *Journal of Political Philosophy* 18: 389–404.

Fung, Archon, and Dara O'Rourke. 2000. 'Reinventing Environmental Regulation from the Grassroots Up: Explaining and Expanding on the Success of the Toxics Release Inventory'. *Environmental Management* 25: 115–27.

Gardiner, Stephen M. 2010. 'Is 'Arming the Future' with Geoengineering Really the Lesser Evil?' In *Climate Ethics: Essential Readings*, edited by Gardiner, Simon Caney, Dale Jamieson, and Henry Shue. New York: Oxford University Press, 284–312.

———. 2011. *A Perfect Moral Storm: The Ethical Tragedy of Climate Change*. New York: Oxford University Press.

Gupta, Aarti. 2010. 'Transparency as Contested Political Terrain: Who Knows What about the Global GMO Trade and Why Does It Matter?' *Global Environmental Politics* 10: 32–52.

Harrison, Kathryn, and Werner Antweiler. 2003. 'Incentives for Pollution Abatement: Regulation, Regulatory Threats, and Non-Governmental Pressures'. *Journal of Policy Analysis and Management* 22: 361–82.

Hayward, Tim. 2006. 'Global Justice and the Distribution of Natural Resources'. *Political Studies* 54: 349–69.

Hiller, Avram. 2011. 'Climate Change and Individual Responsibility'. *The Monist* 94: 349-68.

Jaffee, Daniel. 2012. 'Weak Coffee: Certification and Co-optation in the Fair Trade Movement'. *Social Problems* 59: 94–116.

Kaiser, Michel J., and Gareth Edwards-Jones. 2006. 'The Role of Ecolabeling in Fisheries Management and Conservation'. *Conservation Biology* 20: 392–98.

Keohane, Robert O. 2006. 'Accountability in World Politics'. *Scandinavian Political Studies* 29: 75–87.

Kutz, Christopher. 2000. *Complicity: Ethics and Law for a Collective Age*. New York: Cambridge University Press.

Langley, Paul. 2001. 'Transparency in the Making of Global Environmental Governance'. *Global Society* 15: 73–92.

Levi, Margaret, and April Linton. 2003. 'Fair Trade: One Cup at a Time?' *Politics & Society* 31: 407–32.

Miller, Avram. 2011. 'Climate Change and Individual Responsibility'. *The Monist* 94: 349–68.

Mitchell, Ronald B. 1998. 'Sources of Transparency: Information Systems in International Regimes'. *International Studies Quarterly* 42: 109–30.

———. 2011. 'Transparency for Governance: The Mechanisms and Effectiveness of Disclosure-based and Education-based Transparency Policies'. *Ecological Economics* 70: 1882–90.

Mol, Arthur P.J. 2008. *Environmental Reform in the Information Age: The Contours of Informational Governance*. New York: Cambridge University Press.

———. 2010. 'The Future of Transparency: Power, Pitfalls, and Promises'. *Global Environmental Politics* 10: 132–43

Nolt, John. 2011. 'How Harmful are the Average American's Greenhouse Gas Emissions?' *Ethics, Policy & Environment* 14: 3–10.

Orts, Eric W. 1995. 'A Reflexive Model of Environmental Regulation'. *Business Ethics Quarterly* 5: 779–94.

Philpott, S.M., P. Michier, R. Rice, and R. Greenberg. 2007. 'Field-testing Ecological and Economic Benefits of Coffee Certification Programs'. *Conservation Biology* 21: 975–85.

Ramkumar, Vivek, and Elena Petkova. 2007. 'Transparency and Environmental Governance'. In *The right to know: Transparency for an open world*, edited by Ann Florini, 279–308. New York: Columbia University Press.

Sabel, Charles, Archon Fung, and Bradley Karkkainen. 1999. 'Beyond Backyard Environmentalism'. *Boston Review* (online edition). http://bostonreview.net/forum/beyond-backyard-environmentalism/charles-sabel-archon-fung-bradley-karkkainen-sabel-fung-and.

Stephan, Mark, 2003. 'Environmental Information Disclosure Systems: They Work, but Why?' *Social Science Quarterly* 83: 190–205.

Szasz, Andrew. 2007. *Shopping Our Way to Safety: How We Changed from Protecting the Environment to Protecting Ourselves*. Minneapolis, MN: University of Minnesota Press.

Thaler, Richard H., and Cass R. Sunstein. 2009. *Nudge: Improving Decisions about Health, Wealth, and Happiness*. New York: Penguin.

Tobler, Christina, Vivienne, Visschers and Michael Siegrist. 2011. 'Eating Green: Consumers' Willingness to Adopt Ecological Food Consumption Behaviours', *Appetite* 57: 674–82.

Vandenbergh, Michael, and Anne Steinemann. 2007. 'The Carbon-Neutral Individual'. *New York University Law Review* 82: 1673–1745.

Vanderheiden, Steve. 2009. 'Allocating Ecological Space'. *Journal of Social Philosophy* 40: 257–75.

———. 2011. 'Globalizing Responsibility for Climate Change'. *Ethics & International Affairs* 25: 65–84.

Weil, David, Archon Fung, Mary Graham, and Elena Fagotto. 2006. 'The Effectiveness of Regulatory Disclosure Policies'. *Journal of Policy Analysis and Management* 25: 155–81.

Weins, David. 2012. 'Prescribing Institutions without Ideal Theory'. *Journal of Political Philosophy* 20: 45–70.

Young, Iris Marion. 2004. 'Responsibility and Global Labor Justice'. *Journal of Political Philosophy* 12: 365–88.

*IV*

# New Technologies for Climate Crisis

*Chapter Seven*

# Is There Anything New Under the Sun?[1]

## *Exceptionalism, Novelty and Debating Geoengineering Governance*

Clare Heyward

## INTRODUCTION

'Geoengineering', also known as 'climate engineering', is a term used to cover a wide range of proposals for technologies that, in some sense, will constitute a 'deliberate large-scale manipulation of the Earth's planetary systems in order to counteract anthropogenic climate change' (Shepherd et al. 2009, 1). Some scientists, scientific bodies and other political organisations argue that there should be increased research into geoengineering technologies.[2] There are several justifications offered as to why research into geoengineering should be pursued (see Rayner et al. 2013, 501). The majority of them are motivated by the belief that reductions in global greenhouse gas emissions to date have not been sufficient to give humanity a greater than 50 percent chance of limiting average global temperature rise to 2°C above pre-industrial temperatures plus pessimism that sufficient international co-operation on mitigation will be forthcoming in the near future. The call for increased research into geoengineering technologies has been accompanied by calls for social science and humanities research into the legal, social, political and ethical issues that the technologies might pose. Like any new technology, the research and development of geoengineering technologies raises questions about how eventual use of that technology is going to be controlled, constrained and by whom. Who decides when a geoengineering technology may be used, and upon what grounds may they base their decision?

What measures are appropriate to limit risks and to compensate any harms that do occur?[3] Similar questions apply to the process of research.

Most commentators agree that there are significant governance gaps. There is not, and there is not likely to be, any single mechanism that will be able to regulate all the different technologies that are discussed under the rubric of geoengineering, due to the sheer diversity of the category. Conventionally, geoengineering technologies are categorised as carbon dioxide removal (CDR) technologies, or solar radiation management (SRM) technologies. CDR technologies aim to ameliorate anthropogenic climate change by reducing atmospheric carbon dioxide. SRM technologies aim to reduce temperature increase by increasing albedo (reflectivity) so that less solar energy becomes absorbed by atmospheric greenhouse gases. However the CDR/SRM distinction does not capture the ways in which different geoengineering technologies intervene in different environmental processes and ecosystems and operate at different sites and scales. Land-based technologies, such as BECCS (bioenergy and carbon capture and sequestration) and direct air capture, potentially compete with other proposals for land use and 'justice in siting', but, unless placed near national borders, they at least might come under national jurisdictions, as would settlement brightening. Local planning laws would be relevant for those land-based techniques. Others, such as the sea-based techniques of ocean fertilisation and marine cloud brightening, and atmosphere-based techniques, such as stratospheric sulphate aerosol injection additionally raise questions of transboundary regulation. Examples of relevant international laws and treaties include:[4]

1. The 1972 London Convention on the Prevention of Marine Pollution by Dumping of Wastes and other Matter and 1996 Protocol (LC/LP). Early field trials of ocean iron fertilization led to a 'statement of concern' being endorsed by the Parties to the Convention, in 2007. A year later, the Parties adopted a resolution agreeing that ocean fertilization is governed by the treaty but that 'legitimate scientific research' does not count as dumping, thus raising the question of what counts as 'legitimate scientific research'.
2. The 1977 Convention on the Prohibition of Military or Any Other Hostile Use of Environmental Modification Techniques Convention Environmental Modification (ENMOD). ENMOD prohibits military or any other hostile uses of environmental modification having widespread, long-lasting or severe effects. It does however, permit peaceful use and so does not give positive guidance on when a geoengineering technique might be used.
3. The 1985 Vienna Convention for the Protection of the Ozone Layer. A possible side effect of sulphate aerosol injection (SAI) is to increase ozone depletion, which could constitute a breach of the obligation to

protect the environment against human activities which modify, or are likely to modify, the ozone layer. However, this obligation is based on the desire to avoid 'adverse effects', or 'potential impacts' on human health, which raises the question of how these are to be specified.

4. The 1992 Convention on Biological Diversity. In 2010, Parties to the CDB stated that 'no climate-related geo-engineering activities that may affect biodiversity take place, until there is an adequate scientific basis on which to justify such activities and appropriate consideration of the associated risks for the environment and biodiversity and associated social, economic and cultural impacts'. They made an exception for 'small scale scientific research studies', provided they were subject to oversight.

Thus whilst there are some instruments that could be relevant, there needs to be innovation and co-ordination to ensure that geoengineering technologies are governed appropriately. However, I am not going to offer a specific proposal in this piece. My concern is on what I take to be a precondition of suitable governance mechanisms: a sensible and constructive debate about the place of geoengineering technologies in a portfolio of responses to climate change, and their respective merits and challenges. I shall focus on one possible obstacle to such a debate: the presence of *claims of exceptionalism* that have appeared in the recent debates about geoengineering. A claim of exceptionalism posits that the development of geoengineering will lead to unprecedented situations, either good or bad. Such claims have been made by those who strongly advocate technological research and those who profess deep worries about the consequences of further research (even if they do not oppose it outright). In the geoengineering discourse, it is sulphate aerosol injection that is usually the subject of exceptionalist claims.

The concern expressed in this chapter is that claims of exceptionalism are not conducive to constructive debate on the place (if any) of geoengineering technologies in societal responses to climate change. This is firstly because they are simplistic which I try to show in sections 3 and 4, and secondly (and relatedly), they have a polarising effect. Section 5 notes that the geoengineering debate might be following a pattern of other new technologies; a pattern that Steve Rayner calls the 'novelty trap'. In order to have a productive and constructive debate, I suggest later in the chapter that participants in debates about geoengineering refrain from using exceptionalist claims or if they have previously made them, then to retreat from exceptionalism. However, any retreat must be done in a way which avoids substituting similarly simplistic claims of exceptionalism for claims of continuity.

## EXCEPTIONALISM AND SAI: SOME PRELIMINARIES

As this chapter is about claims of exceptionalism and novelty, I should begin by pointing out that I am not the only one, much less the first, to highlight that a proportion of the discussions about geoengineering invoke such claims. For example, Tina Sikka identified a 'rhetoric of exceptionalism' (Sikka 2012, 68) in support for geoengineering research. Sikka defines exceptionalism as 'the process of setting up often false, ominous *and therefore exceptional*, scenarios in which we as citizens must choose between two stark and generally unappealing choices'. However, I would like to offer a broader understanding of 'exceptionalist claims' which takes account of the idea that things can be exceptionally good or exceptionally bad. Let us say, firstly, that claims of exceptionalism are simply those which posit extraordinary circumstances. Secondly, a putative analytic distinction can be made between:

1. issues arising from features of the geoengineering technology that might pose a challenge for governance systems; and
2. issues that might arise in creating and maintaining (legitimate) governance systems for geoengineering technologies.

In category (1), claims of exceptionalism are those that make explicit or implicit reference to unprecedented situations that are favourable or adverse. For example, some have claimed that one geoengineering technique, SAI, is a *very* cheap way of combating temperature rises and that this is a reason to pursue it (see for example Barrett 2008; Bickel and Lane 2009; Weitzman 2011). The exceptionalist element to this claim is positive. It is not simply that SAI could keep global temperatures at a desirable level, which would mean that the problems of distributive justice associated with climate change would be ameliorated, but *that it could do so at a very low cost* (e.g., compared to mitigation). Call this, following Scott Barrett, the 'incredible economics' argument.[5] An example of a negative exceptionalist claim is the objection concerning the 'termination effect'. Like many SRM methods, if SAI is used without reducing GHG emissions, and for any reason the technology is 'switched off', there will be a rapid temperature increase to which it will be very hard to adapt. The concern about the termination effect is thus about the possible risk of a new and extremely adverse predicament.

Another kind of exceptionalist claim, which was especially prominent at the start of the debate, is the 'climate emergency justification' of research into SAI. As concerns about the possible passing of 'climate tipping points' became more frequently expressed, some suggested that SAI would be able to quickly bring down global temperatures in order to avert a so-called climate emergency.[6] The climate emergency argument for SAI research posited new and adverse circumstances and suggested that SAI could be a solution. Like the 'incredible economics' justification, it is ultimately a positive claim. However, it is one that gains its rhetorical power from the idea that an extremely adverse situation might arise, meaning that the great benefit posited is the avoidance of a great harm.

Turn now to category (2). Rather than looking at the issues that governance systems involving geoengineering technologies might have to deal with when they are created, the second category of 'geoengineering governance challenges' is that of creating and/or maintaining those systems of governance, whether they be formal or informal. Challenges in category (2) might raise questions of practicality and also of legitimacy. For an example of a practical challenge, it might be argued that research or deployment of geoengineering would be difficult, even impossible to govern (Hulme 2014, 57). Another example is the argument against a moratorium on geoengineering research on the grounds that it would not stop the less scrupulous from pursing such research (e.g., Parker 2014, 6).

Challenges of legitimacy are a cluster of concerns about how decision-making processes concerning geoengineering technologies might be manifested. Several have warned that commercial or other vested interests might come to dominate decisions about which geoengineering technologies to pursue, or whether to deploy (Hamilton 2013; Szersynski et al. 2013, 2813). Others have warned that there is 'a nearly inexorable trajectory from research to deployment' (Jamieson 1996, 333). This objection, often referred to as the 'slippery slope' (e.g., Irfan 2013), postulates that the power of any small minority increases as research proceeds, thus generating its own momentum. The overall concern is that the broader public will not have an adequate say in what kinds of geoengineering technologies are acceptable to research or to use.

Claims of exceptionalism in category (2) are those that make explicit or implicit claims that geoengineering technologies will create unprecedented challenges for legitimate governance. Unlike the first category, where claims have been made with regard to positive and negative effects, to date, those in category (2) have been overwhelmingly negative. For example, some claim that SAI might be used unilaterally by a state (Victor et al. 2013), or a 'Greenfinger' (Victor 2008). A single agent being able to affect global temperatures so greatly would be an unprecedented and unwelcome situation. As noted already, Hulme (2014, 57) has argued that SAI is 'ungovernable', or rather, ungovernable by current democratic mechanisms. Similarly, according to Szersysnki et al. (2013), 'SRM [sulphate aerosol injection] and democracy won't mix'. SAI, they argue, generates 'an unprecedented set of challenges for democracy' (2810). This is due to:

1. 'inherent uncertainties',

2. 'acting at the planetary level and necessitating autocratic governance',
3. 'plural and unstable motivations', and
4. 'becoming conditioned by economic forces'.

Some of these reasons, namely (2) and (3), invoke claims of exceptionalism. We shall consider those arguments in the next section.

## Exceptional Governance Challenges of SAI?

As we saw above, others have described SAI as being 'ungovernable' (Hulme 2014), or as potentially 'negating' democracy (Macnaghten & Szerszynski 2013, 472). We shall now look at some reasons given for doubting the compatibility of SAI and democratic governance and see that they do not yet make a fully convincing case.[7]

One reason for thinking that governance of SAI 'will not mix' with democracy (Szerszynski et al. 2013) arises from the claim that its effects are manifest on a planetary scale and that it must be controlled centrally, presumably by a global body.[8] Macnaghten and Szerszynski express this idea as follows:

> Democracy, in its various forms, depends on the articulation, negotiation and accommodation of plural views and interests. It relies on an evolving and partially flexible relationship between citizens and governance institutions. Solar radiation management [sic] *by contrast exists as a planetary technology*. While plausibly able to accommodate diverse views into the formulation of its use, once deployed, *there remains little opportunity for opt-out or for the accommodation of diverse perspectives*. By its social constitution it appears inimical to the accommodation of difference. Following deployment it could only be controlled centrally and on a planetary scale. (2013, 472, emphasis added)

Macnaghten and Szerszynski seem to claim that governance of SAI must take place on a planetary scale, as using SAI will have effects on the global climate. We might also discern a more specific claim: that democratic governance requires that individual members are able either to determine the outcome of a decision-making process, that is, to have their preferences satisfied or to opt out of a decision should that decision run counter to their wishes. Macnaghten and Szerszynski seem to imply that because SAI would affect the global climate, an agent cannot opt out from, or escape the impacts of, SAI if it were implemented.

Is it plausible that to count as democratic, that individuals (persons or states) have a right to not only express dissent, but to determine whether SAI is used? Any decision-making process which requires a consensus from all participants (such as that of The Conference of the Parties to the United

Nations Framework Convention on Climate Change) effectively grants each participant the ability to veto a proposed course of action and thus to determine the outcome. However, within societies that are uncontroversially deemed democratic, a form of majoritarian democracy is the most common. Under this system, participants may influence the decision by casting their vote, but they do not fully determine the outcome. Nor, under this system, do participants have the right to 'opt out' if the outcome is contrary to their preferences. Citizens of democratic countries are obligated to obey the laws enacted democratically, even if those laws run contrary to their reasonable beliefs about politics (there are exceptions if the society starts to commit severe injustices). A lack of ability to opt out does not, by itself render a process undemocratic. Without further elaboration, therefore, a lack of an opt-out does not show that SAI and democracy are incompatible.

As the world grows increasingly interconnected, there are calls for greater global governance and accompanying fears that global political institutions would be tyrannical. Perhaps Szersynski and his colleagues are restating this perennial fear. For example, a common objection to the idea of a world state is that there would be no escape should it turn out to be tyrannical. The lack of viable exit or opt-out does not, however, render a world state undemocratic (although it does, arguably, make it more important that democracy is achieved).

We also might draw a parallel with an argument made by the International Relations theorist Terry Nardin (1983, 238–9). Nardin argues that individual freedom is threatened when power is centralised, and best safeguarded in governance systems with checks and balances. He thought that this is best maintained in a world with plural jurisdictions (i.e., where many states exhibit sovereignty). However, there are responses available. First, and most obviously, there is a difference between a world government/world state and global governance. There is some work to do in showing that the issues with the former also apply to the latter. Let us assume for the moment that the worry is that an institution governing SAI would have to be global in scope and that there would have to be some concentration of power (this usually is the case if the institution is worthy of the name). It is conceptually plausible that such a global institution could have checks on its power, or be balanced by other global institutions. There are proposals for cosmpolitan democracies which feature plural, overlapping global institutions. It might be possible that institutions governing SAI can be part of such multi-lateral global orders. Much would have to be done to establish appropriate checks and balances, and maybe it would turn out to be impracticable. However, it is not enough merely to assume or assert that global or centralised political structures must be undemocratic.

Mike Hulme also claims that SAI would be 'ungovernable' in a democratic manner. He is perhaps best interpreted as making a claim about the likelihood that such a cosmopolitan system could emerge, given the current international order. Governing SAI would require a huge degree of global co-operation, which Hulme doubts will happen (Hulme 2014, 49). It is also likely that there will be disputes about whether to implement SAI and perfectly possible that it could open up a new channel for inter-state rivalry and disagreement (Hulme 2014, 48). Such scepticism about the prospects of governing SAI is understandable but not final word.

Note that I do not claim that SAI, if successfully developed, *will* be governed by a democratic system. There are serious challenges of governance of SAI, but it is one thing to adumbrate the challenges and another to say that they are unsolvable. I am simply trying to show that much more needs to be said about the possibilities of global governance and conceptions of democracy before we can conclude that SAI is incompatible with, or raises unprecedented challenges for democracy.

Another reason given as to why SAI raises unprecedented challenges for democracy refers not to the global nature of SAI interventions but to what Szerszynski and various colleagues refer to as the problem of 'plurality of intent'. The concern seems to stem from the thought that there might be several motivations for undertaking a programme of SAI. Szerszynski and his colleagues claim that this creates a problem because SAI, as an example of geoengineering, is defined as *deliberate intervention in order to counteract anthropogenic climate change*. Therefore, they claim, whether or not an action counts as geoengineering deployment, as research or even as mere pollution will be settled according to its purpose. They state 'this means that meaningful geoengineering governance would logically require the scrutiny and regulation of the intentions, whether explicit or implicit, of a huge range of research and deployment activities, raising significant logistical and political questions'. They go on to assert that due to there being different motivations for undertaking SAI, the relationship between intention, deployment and consequences will 'always be unstable, rendering problematic any claims of prior democratic consent and leading to new kinds of conflict and controversy' (Szerszynski et al. 2013, 2813).

Unfortunately, the authors do not give examples either of the significant logistical and political questions, or the new kinds of controversy. Regarding the former, we might imagine one such problem as follows: in order to govern anything, there has to be a fairly clear idea of the object(s) to be governed. If geoengineering is distinguished from other activities only by agents' intent, then it will be difficult to establish what activities should fall under the remit of a geoengineering governance scheme. As an example, consider the project E-PEACE, the Eastern Pacific Emitted Aerosol Cloud Experiment, based at the University of California at San Diego, which is researching aerosol-cloud dynamics, in particular the effects of interactions between aerosols and clouds on the reflection of solar radiation by clouds.

The E-PEACE researchers present their work as increasing the understanding of clouds and hence contributing to climate science generally.[9] Increasing the reflectivity of clouds, however, is perhaps the second-most talked about SRM technology, under the name of 'marine cloud brightening'. Had they chosen to, these researchers could have highlighted the possibility of understanding cloud reflectivity *in order to affect it* and thus as contributing to geoengineering research. Or consider the more widely reported Haida Gwaii Salmon Restoration (HGSR) project, which added iron nutrients to the ocean, bringing about an algal bloom. The project was billed as an attempt to restore salmon to that region of the sea traditionally fished by the indigenous Haida Gwaii community. However, some directors of the company behind the project, in particular an American businessman, Russ George, were also interested in the possibility of using the plankton to sequester carbon dioxide. Thus the project could either be cast as attempting to bring back salmon, or as a test of ocean iron fertilisation, a form of carbon dioxide removal.

It might seem that any system of geoengineering governance would have to examine projects such as these and adjudicate whether the main investigators were ultimately interested in developing ways to increase cloud albedo and carbon dioxide sequestration, or simply interested in cloud physics and restoring fishing prospects. However, this need not be the case. What is important is that *technologies and activities currently described under the term geoengineering*, including SAI, are adequately governed to ensure that they do not severely harm environmental systems and/or cause or exacerbate poverty, land dispossession, or other injustices. If a technology or activity is considered to (potentially) have adverse consequences, then it should come under some sort of governance regime, regardless of whether it is being pursued in order to counteract anthropogenic climate change or not. Many systems of governance focus on activities and situations. Ascertaining purposes and intentions might be important *within* these systems, but not in defining their remit.

Return to the example of the salmon restoration/ocean fertilisation project. The London Protocol prohibits the dumping of materials in the oceans (with some exceptions). There is good reason to restrict the ingress of materials into the oceans because of their effects on marine ecosystems. *Activities that release materials into the oceans* thus come under a system of governance. In one important respect it did not matter whether the HGSR project was an attempt at restoring salmon or an ocean iron fertilisation experiment. It involved the release of active, inorganic particles into ocean waters and thus it came under certain governance arrangements. To take an even more controversial example, the practice of whaling is monitored by the International Whaling Commission (IWC). Activities that involve the killing of whales come under its remit.

In 1986 the IWC imposed a moratorium on commercial whaling, but subsistence whaling by indigenous communities and 'scientific' whaling is still permitted. The *justification* of an activity often invokes its purpose. It is permissible to hunt whales for subsistence purposes or for scientific research, but not for any other purpose. But it is the *activity* 'hunting whales' that determines the remit of the IWC. It is then up to that institution to determine catch limits and it does so by setting different catch limits to different groups, who purport to achieve different purposes in hunting whales. It can, and indeed has been debated as to what purposes justify the taking of whales. Also, it can, and has been debated whether certain groups are making misleading statements about their purposes. For example, Japanese claims that their whaling was for scientific purposes have long been contested. This controversial issue was brought to the International Court of Justice, which ruled earlier this year that the Japanese claim to be engaged in scientific whaling was unsupported and ordered Japan to cease its whaling programme.[10] Szersynski and his colleagues are thus correct that intentions and purposes matter and can cause controversy and conflict. But they need not matter so much when determining the remit of governance systems. Moreover, as the whaling example shows, governance and regulatory systems can and do regularly adjudicate on the presence of intentions. Wherever questions of fault arise, so do issues of intent.

Nor do Szerszynski et al. indicate how the 'instability of intent' might render 'problematic any claims of prior democratic consent'. One way of understanding this concern is that SAI might become used in ways other than the original purpose of lowering average global temperature. Without further explanation, this claim does appear questionable. The authors state that SAI might come to be regarded as a means of achieving 'humanitarian, environmental, nationalistic, military or commercial goals' (Szerszynski et al. 2013, 2813). However, SAI could only serve those goals by increasing albedo. As such, it could still be defined by deliberately intervening in the global climate in order to counteract anthropogenic climate change. The wider purpose of SAI might be changed slightly *to further goal X by counteracting anthropogenic climate change*, but this does not appear to be a major difference. Moreover, if we consider the first two 'alternative' goals of humanitarianism and environmentalism, they do not appear hugely different to the goal of counteracting climate change. Very few, if any, argue that climate change should be limited for its own sake. Anthropogenic climate change is a matter of concern because of the impacts it is projected to have on the poorer regions of the world, as well as upon ecosystems and biodiversity. Action on climate change gains its primary and most compelling justification on humanitarian (or global justice) and environmental grounds. There are also economic cases for addressing climate change (Stern 2007). As for 'nationalistic' or 'military' goals, much more needs to be said (1) about what they are and (2) to show that SAI is an effective means of furthering them. There are many quicker and more direct ways of achieving military aims, i.e., many more effective weapons than reducing the global mean temperature by a degree or so. Again, if by 'nationalistic' purposes, we understand the increasing a nation-state's power vis-à-vis its neighbours or the global community, then again, it seems that there are quicker and more direct routes than affecting the climate.

Regardless of the above, however, it is simply unclear at least without further explanation why the 'plurality of intent' poses a problem for democratic governance. Citizens can agree to policies and actions for different reasons (or different weighing of reasons). Note however that the precise claim is that the instability of motivation renders problematic '*prior* democratic consent'. The thought might be that a democratic body might agree to using SAI for one specific purpose, but it is then used to achieve a different goal, which the citizenry did not initially 'sign up to'. This could pose a problem, but does not seem to be a particularly rare or insurmountable one. Indeed, political goals and the instruments, mechanisms and technologies to achieve them should be scrutinised and subject to change in the course of democratic politics. Moreover, what appears to be of concern here is not simply that the motivations for using SAI might be plural, but a more general concern that a polity might adopt undesirable or unjust goals. If this is the real concern, then I agree it is serious. This a problem faced by all societies, including democratic ones (and not only in the case of SAI). Thus further argument is needed to establish that the existence of plural motivations calls into question whether SAI can be democratically governed.

However, it is unsurprising that there has been a negative reaction to SAI from these authors and others. Many new technologies raise questions of trust, liability and consent (Rayner and Cantor 1987). Additionally, as I shall suggest in the next section the 'climate emergency justification' primed audiences to ask questions about social benefits and legitimacy.

## EXCEPTIONALISM IN SAI RESEARCH ADVOCACY

The idea that geoengineering could be required to deal with a future 'climate emergency' became a prominent theme in many early articles and reports advocating research efforts into geoengineering. Paul Crutzen, whose 2006 article is widely credited with 'breaking the taboo' on geoengineering research, quoted the eminent climate scientist Stephen Schneider: 'supposing a currently envisioned low probability but high consequence outcome really started to unfold in the decades ahead (for example 5°C warming this century, which I would characterize as having potentially catastrophic implications for ecosystems … Under such a scenario, we would simply have to practice geoengineering' (Schneider 1996, quoted by Crutzen 2006, 214). Following Crutzen, many argued that there was a pressing need for research into SAI because it promises to be a fast-acting technology that could be deployed to avert abrupt climate events (Blackstock et al. 2009; Victor et al. 2009; Caldeira and Keith 2010; Long et al. 2011; Blackstock and Long 2011; Goldblatt and Watson 2012; Victor et al. 2013).

Arguably, it was the idea of a climate emergency that broke the scientific community's self-imposed taboo on the advocacy of geoengineering research (forthcoming Heyward and Rayner 2015). It did so by emphasising the exceptional, unprecedented circumstances of the 'climate emergency'.

Thus, whilst these and many other researchers professed reluctance about the prospects of SAI, emphasising that geoengineering with SAI should not normally be considered (i.e., it should not be regarded as a substitute for the mitigation), they nevertheless argued that a new and extremely serious problem, the climate emergency, warranted a new approach to dealing with climate change.

The climate emergency justification quickly came in for criticism. One problem was its vagueness. It was rare for there to be a sustained discussion of what a climate emergency might be, and even rarer for there to be any discussion on its likelihood.[11] For example, a report on the potential of geoengineering technologies gave a simple, formal definition: a climate emergency is where a serious impact would occur too rapidly to be averted by even immediate mitigation efforts (Blackstock et al. 2009, 1). Following other examples, (e.g., Victor et al. 1996), the report later gave some of the common examples of abrupt climate change and tipping points, then went on to state:

> While offering no predictions about the likelihood of either climate emergencies or gradual climate change (or anything in between), we recognize that there is now a non-negligible possibility of a climate emergency in one form or other. (Blackstock et al. 2009, 22)

The positing of a relatively undefined 'climate emergency' thus invited the question of what might count as a climate emergency. Without recourse to a specific definition, the climate emergency argument appeared to invoke a vague threat and thus vague benefit to justify a course of action. It is understandable that this should invite questions. It is far from clear that SAI can be a plausible (let alone the best or only) response to a climate emergency until we know what such a climate emergency is.

Even if there is agreement on what would constitute a true 'climate emergency' then it cannot be simply assumed that use of SAI is the best, let alone the only response. Sometimes solutions to a serious problem are themselves morally unacceptable. Particularly pertinent here is that the declaration of an emergency is often used to justify a certain response, which involves a suspension of normal rules, and often the suppression of dissent. Therefore, the invocation of 'emergency' might well have prompted questions on what kinds of measures technical and political might be deemed justified if a climate emergency was declared.

The emergency rhetoric itself highlighted potential problems outlined above: the vagueness of the term 'climate emergency', plus the associations between emergency rhetoric and suspension of normal mechanisms of governance. Thus, it is not surprising that such questions were seized upon by those less enthusisatic about the prospect of researching SAI.

## The Novelty Trap?

Writing a decade ago when nanotechnology was trumpeted as the smallest ever 'next big thing', Steve Rayner observed that the then emerging debate appeared to develop the same way as previous disputes about new technologies. He termed this the 'novelty trap' (Rayner 2004). The pattern is as follows: advocates and enthusiasts emphasise the power and potential of the technology in question and promise that it will usher in a period of immense prosperity, or otherwise make the world immeasurably better. In response 'sceptics are wont to question the idea of a free, or at least almost free, all-you-can-eat lunch. If the product or process is so new and revolutionary, they argue, can there also be new and revolutionary drawbacks?' (Rayner 2004, 351).

In the face of such concerns, advocates of the technology emphasise its similarities with familiar processes and activities. The once-revolutionary technology is, all of a sudden, another way of doing something humanity has been doing for years. Rayner gives examples including: nuclear reactors being described as 'just another way of boiling water', GMOs being described as 'extensions of plant-breeding', and assurances that nanotechnology is 'just a new name for chemistry or materials science and, to prove it, we are reminded of the colloidal gold nanoparticles produced by Michael Faraday in the 1850s' (Rayner 2004, 352).

Here we might ask: Is the debate about geoengineering approaching the novelty trap? The climate emergency justification for SAI development pointed to the potential of great, but vaguely defined, benefits (or rather the avoidance of vaguely defined harms that *might* happen). The 'incredible economics' argument also claimed great benefits. In response, there have been questions about the potential costs of side effects of SAI and fears expressed (although in an imprecise way) that SAI raises problems for democratic goverance that *might* be insurmountable. Thus we can see that a proportion of those who support SAI research and a proportion of those who are sceptical of SAI's place in societies' responses to climate change have both made claims which (1) make reference to exceptional, or unprecedented and novel situations, (2) point to massive benefits or immense problems, but (3) are, I have tried to show, prone to imprecision. We have had some side claiming new and revolutionary benefits, others claiming new and revolutionary drawbacks.

Some prominent advocates of SAI research are rejecting exceptionalism and instead emphasising continuity and familiarity. David Keith, for example, has written a short book advocating SAI research, including some small-scale field trials (Keith 2013). The climate emergency justification is not endorsed and Keith is critical of those who present climate change as being a matter of imminent catastrophe (Keith 2013, 24). Elsewhere, he has been reported as saying that 'framing the case for geoengineering as a last resort in a climate emergency is 'a bit of a rhetorical trick': it leaves undefined what a 'climate emergency' is, and 'there is no simple definition'. (David Keith, quoted by Rotmans 2013).

Rather, Keith has started to downplay novelty. Obviously, humans have not tried to address climate change by SAI before, but, Keith says, sulphur dioxide and sulphate aerosols have been studied and used before. He writes: 'We are not leaping into the unknown with a totally new chemical. Nor are the risks irreversible as in, for example, releasing a species with novel engineered genes' (Keith 2013, 12). Indeed, every year, the world produces 50 million tonnes of sulphate emissions. Note that Keith is not saying that the practice of releasing sulphates into the atmosphere was a good thing. Rather, he is suggesting *familiarity*: he is suggesting that our previous experience in monitoring and dealing with sulphur-based air pollution (usually seen to be a problem) justifies some confidence that SAI and its effects can be monitored and managed.

Ken Caldeira, another leading advocate and researcher of SAI, has recently voiced some disquiet about 'geoengineering exceptionalism': the idea that 'geoengineering research ought to be treated differently from other kinds of scientific research' (posted on Googlegroups, 21 August 2014). In a recent conference plenary discussion, Caldeira's chosen theme was: what's special about climate engineering research or climate engineering?' He asked, rhetorically:

> Does it make any sense to have a meeting that discusses BECCS [bioenergy, carbon capture and sequestration] that does not also discuss biomass energy, that doesn't discuss CCS [carbon capture and sequestration]? As soon as we lump these two together, suddenly it becomes climate engineering and we're supposed to worry about its implications. If I have two power plants: a coal plant, a biomass energy plant [and] a CCS facility. If the pipe goes from the coal plant to the CCS plant and the biomass energy goes off [gestures upwards]. That's all mitigation, fine, no problem, no ethical issues that we need to have big conferences about. I'm exaggerating of course. But as soon as you

> switch that pipe so that now the biomass facility is going to the CCS facility and the coal plant is emitting to the atmosphere, it's become climate engineering and ... because it's climate engineering we're supposed to treat it specially (Calderia, speaking at CEC14 panel 'From Fringe to Fashion? The Past Decade of Climate Engineering Research').[12]

Caldeira's claim seems to be that there is nothing about geoengineering research, or activities described as geoengineering that need cause additional concern or scrutiny. Other researchers during that conference session echoed that theme, e.g., asking whether there was any practical difference between the E-PEACE experiment and 'geoengineering research'.

So we can see that at least some who research SAI are withdrawing from making exceptionalist claims and instead challenging them. At the same time, continuity is emphasised, in this case with other research on climate change and other ways of responding to climate change. Caldeira could be understood as implying that it is unfair to researchers to put certain forms of geoengineering reseach under additional scrutiny because there is no signficant difference between it and other scientific research.

## ENDING EXCEPTIONALISM: TOWARDS A MORE CONSTRUCTIVE DEBATE?

Climate change is a very serious problem, which requires a response. For that a reasoned and constructive discussion is needed.

Exceptionalist claims have a tendency to bring about polarisation of the issues, which encourages a 'with us' or 'against us' mentality. The strong but imprecise claims characteristic of exceptionalism in the SAI debate are not only easily dismissed, but can also give the impression that one is searching for an excuse to do what one already wants to do (if an enthusiast or advocate), or if sceptical, that one's scepticism is simply an attempt to vocalise an inchoate gut reaction against technology. Once this has happened, subsequent reasonable claims, positive or negative might be dismissed because they are seen to emerge from gung-ho advocates or modern-day Luddites. Moreover, switching from claims of novelty to claims of continuity risks a loss of credibility, since it will be seen as a case of 'over-claiming benefits and back-pedalling in the face of criticism' (Rayner 2004, 352). This might be especially so if the claims are simplistic, as the switch might be regarded as a PR exercise, rather than the outcome of a genuine intellectual process.

One answer is for those involved in the geoengineering debates to refrain from making claims of exceptionalism. As we saw in the previous section, some are beginning to do so. However, it is vital that claims of exceptionalism are not replaced with simplistic claims of continuity. Rather, the complexity of the various kinds of questions raised by CDR and SRM technologies should be acknowledged. In particular, it should be acknowledged that one of the complexities is the extent to which any geoengineering technology should be pursued at all. To complain, as Caldeira seems to, about 'extra scrutiny' of projects labelled as 'geoengineering' is in some sense to miss the point. If the case for researching a form of technology is justified according to its potential impact on climate change, questions not only about its safety etc., but about its benefits and disadvantages compared to other potential responses to climate change are to be expected. It is a reasonable question to ask whether there should be a 'Manhattan Project' for SAI (as recommended by Jay Michaelson (1998), or whether similar or greater efforts should be made into researching clean energy as well as or instead of any geoengineering technology as recommended by some (Gardiner 2010; Hulme, 2014, 85–87). There are questions about CDR and SRM technologies that their advocates face simply because they present those as a means of addressing the problems of climate change. Insofar as that counts as 'extra scrutiny' it is justified and in line with the kind of scrutiny that other responses to climate change have undergone. For example, the proposal to use of biofuels as a form of mitigation has faced questions about the effectiveness of biofuels in relation to other kinds of responses to climate change. It has also been asked whether developing biofuels creates or exacerbates social injustices, e.g. endangering food security by taking over land that was once used for subsistence farming, or displacing people from their lands (see, e.g., Schrader-Frechette, this volume). Similarly, it has long been acknowledged in the mainstream climate discourse that adaptation raises questions of social justice (e.g., Paavola & Adger 2006). Geoengineering technologies should not be discussed in isolation from other kinds of response to climate change (Heyward 2013) and climate change itself should not be discussed in isolation from general concerns of global justice, e.g., food security and global poverty (Caney 2012). As Peter Healey suggests, the goal of good geoengineering technologies assessment and governance should be 'direct, hard-headed comparisons of routes to a variety of environmental food protection and energy goals, centred on, but not confined to climate change objectives' (Healey 2014, 34).

Second, a gradualist approach to governance should be developed. This is necessary not only as an antidote towards exceptionalism, but because developing proposals for the governance of geoengineering technologies face the 'technology control dilemma' (Collingridge, 1980): it is impossible to foretell what any geoengineering proposal now talked about will look like in its fully developed form (assuming it reaches that stage). Adopting a gradualist process helps develop a system of governance in this situation.

One suggestion for a such a gradualist approach is the development of technology-specific research protocols, for every stage of research from the initial idea to any field trials and eventual implementation (Rayner et al.

2013, 509). Rayner et al. argue that before any activity, the researchers should be required to prepare a research protocol explicitly articulating how they are respecting core societal values.[13] This is sent to a review body, which can withhold approval until it is assured that the research design satisfies core societal values and that it will be conscientiously implemented. The precise informational requirements and the identity of the reviewing parties will vary according to the stage of research. National research councils would be an obvious review body in the first instance: if experiments grow bigger in scale, then additional scrutiny will be needed.

Rayner et al. envision that the

> development of technology specific research protocols is the first step of the bottom-up process of building a flexible governance architecture. … [T]hey could serve initially as a code of conduct by scientific researchers and research councils. These more specific regulations generated in the research setting could then be adopted and modified by other institutions, including, where necessary, formal mechanisms such as legal regulation. (Rayner et al. 2013, 509)

Such a gradualist approach is a response to exceptionalism and a rejection of the polarisation that accompanies it. It allows research to proceed, but under scrutiny. The level of scrutiny is, however, appropriate to the stage of development and thus not overly burdensome on the earliest stages of research. The use of stage-gates, whereby permission to continue research can be denied, helps to address concerns. The stage-gate mechanism was used in the only outdoor experiment proposed by the SPICE project, a UK research project looking at the delivery and effects of SAI. The purpose of the experiment was to test a delivery mechanism using water rather than sulphate aerosols, but as it was the first geoengineering-related experiment done in the UK, it was decided that there should be a public engagement process before the test was run. When the review came, the test-bed was postponed in order that further stakeholder engagement could be conducted (Macnaghten & Owen 2011).[14]

This is but one example and much more remains to do in terms of specifying the content of research protocols and the make-up of review-bodies. My point is simply that starting to do so is one possible means of helping to avoid exceptionalist claims, polarised debates, and the novelty trap.

## CONCLUSION

At present it is very uncertain as to how research into geoengineering technologies will proceed and how it will be governed, but much will depend on the quality of debate. It is vital that there is reasoned and constructive debate on the place, if any, of geoengineering technologies in societal responses to climate change. The sooner that simplistic appeals to unprecedented circumstances are dropped, the sooner this debate can take place.

## REFERENCES

Barrett, S. 2008. 'The Incredible Economics of Geoengineering'. *Environmental and Resource Economics* 39, 45–54.

———. 2014. 'Solar Geoengineering's Brave New World: Thoughts on the Governance of an Unprecedented Technology'. *Review of Environmental Economics and Policy* 8, 249–69.

Bickel, E., and L. Lane. 2009. Copenhagen Consensus: An Analysis of Climate Engineering as a Response to Climate Change.

Blackstock, J., D. Battisti, K. Caldeira, K., D. Eardley, J. Katz, D. Keith, A. Patrinos, D. Schrag, R. Socolow, and S. Koonin. 2009. *Climate Engineering Responses to Climate Emergencies*. Santa Barbara, CA: Novim.

Blackstock, J. J., and J. C. S. Long. 2011. 'The Politics of Geoengineering'. *Science* 327, 527.

Bodansky, D. 1996. 'May We Engineer the Climate'? *Climatic Change* 33, 309–21.

———. 2013. 'The Who, What, and Wherefore of Geoengineering Governance'. *Climatic Change* 121, 539–51.

Caldeira, K., and D. Keith. 2010. 'The Need for Geoengineering Research'. *Issues in Science and Technology*, 57–62.

Caney, S. 2012. 'Just Emissions'. *Philosophy & Public Affairs* 40, 255–300.

Cicerone, R. 2006. 'Geoengineering: Encouraging Research and Overseeing Implementation'. *Climatic Change* 77, 221–26.

Collingridge, D. 1980. *The Social Control of Technology*. New York: St Martin's Press.

Crutzen, P. 2006. 'Albedo Enhancement by Stratospheric Sulfur Injections: A Contribution to Resolve a Policy Dilemma'? *Climatic Change* 77, 211–20.

Gardiner, S.M., 2010. 'Is "Arming the Future" with Geoengineering really the Lesser Evil? Some Doubts about Intentionally Manipulating the Climate System'. In Gardiner, S.M., Caney, S., Jamieson, D., Shue, H. (Eds.), *Climate Ethics: Essential Readings*. Oxford: Oxford University Press, 284–312.

Goldblatt, C., and A. J. Watson. 2012. 'The Runaway Greenhouse: Implications for Future Climate Change, Geoengineering and Planetary Atmospheres'. *Philosophical Transactions of the Royal Society A: Mathematical, Physical and Engineering Sciences* 370, 4197–216.

Hamilton, C. 2013. *Earthmasters: The Dawn of the Ages of Climate Engineering*. Princeton, NJ: Yale University Press.

Heyward, C. (2013). 'Situating and Abandoning Geoengineering: A Typology of Five Responses to Dangerous Climate Change' *PS: Political Science and Politics* 46, 23-27.

Heyward, C., and S. S. Rayner 'Apocalypse Nicked!'. In Crate, S. and M. Nuttall (eds) Anthropology and Climate Change 2nd Edition, Left Coast Press, Walnut Creek CA.

Horton, J. 2011. 'Geoengineerng and the Myth of Unilateralism: Pressures and Prospects for International Cooperation'. *Stanford Journal of Law, Science and Policy* 4, 56–69.

Hulme, M. 2014. *Can Science Fix Climate Change: A Case Against Climate Engineering*. London: Polity.

Irfan, U. 2013. 'Would-be Geoengineers Call for Research Guidelines'. *Scientific American*, 18 March 2013, available at http://www.scientificamerican.com/article/would-be-geoengineers-call-for-research-guidelines/

Jamieson, D. 1996. 'Ethics and International Climate Change'. *Climatic Change* 33, 323–36.

Keith, D. 2013. *A Case for Climate Engineering*. Boston: MIT Press.

Keith, D.W. 2000. 'Geoengineering the Climate: History and Prospect'. *Annual Review of Energy and Environment* 25, 245–84.

Long, J., S. Raddekmaker, J. Anderson, R. E. Benedick, K. Caldeira, J. Chaisson, D. Goldston, S. Hamburg, D. Keith, R. Lehman, F. Loy, G. Morgan, D. Sarewitz, T. Schelling, J. Shepherd, D. Victor, D. Whelan, and D. Winickoff. 2011. *Geoengineering: A National Strategic*

*Plan for Research on the Potential Effectiveness, Feasibility and Consequences of Climate Remediation Technologies.* Washington, DC: Bi-Partisan Policy Center.

Macnaghten, P., and R. Owen. 2011. 'Good Governance for Geoengineering'. *Nature* 497, 293.

Macnaghten, P., and B. Szerszynski. 2013. 'Living the Global Social Experiment: An Analysis of Public Discourse on Solar Radiation Management and Its Implications for Governance'. *Global Environmental Change* 23, 465–74.

Michaelson, J. 1998. 'Geoengineering: A Climate Change Manhatten Project'. *Stanford Journal of Environmental Law* 17, 73–40.

Paavola, J., and W. N. Adger. 2006. 'Fair Adaptation to Climate Change'. *Ecological Economics* 56, 594–609.

Parker, A. 2014. 'Governing Solar Geoengineering Research as It Leaves the Laboratory'. *Philosophical Transactions of the Royal Society of London A: Mathematical, Physical and Engineering Sciences*.

Pielke Jr, R. 2010. *The Climate Fix: What Scientists and Politicians Won't Tell You about Global Warming*. New York: Basic Books.

Rayner, S., and R. Cantor (1987) 'How Fair is Safe Enough? The Cultural Approach to Social Technology Choice' *Risk Anaysis* 7, 3-9.

Rayner, S. 2004. 'The Novelty Trap: Why Does Institutional Learning about New Technologies Seem So Difficult'? *Industry and Higher Education* 18, 349–55.

Rayner, S., C. Heyward, T. Kruger, N. Pidgeon, C. Redgwell, and J. Savulescu. 2013. 'The Oxford Principles'. *Climatic Change* 121, 499–512.

Robock, A. 2008. '20 Reasons Why Geoengineering May Be a Bad Idea'. *Bulletin of the Atomic Scientists* 64, 14–18.

Russill, C., and Z. Nyssa, 2009. 'The Tipping Point Trend in Climate Change Communication'. *Global Environmental Change* 19, 336–44.

Schelling, T. 1996. 'The Economic Diplomacy of Geoengineering'. *Climatic Change* 33, 303.

Shepherd, J., K. Caldeira, P. Cox, J. Haigh, D. Keith, B. Launder, G. Mace, G. MacKerron, J. Pyle, S. Rayner, C. Redgwell, and A. Watson. 2009. *Geoengineering the Climate: Science, Governance and Uncertainty.* London: Royal Society.

Sikka, T. 2012. 'A Critical Discourse Analysis of Geoengineering Advocacy'. *Critical Discourse Studies* 9, 163–75.

Stern, N. 2007. *The Economics of Climate Change*, Cambridge: Cambridge University Press.

Szerszynski, B., M. Kearnes, P. Macnaghten, R. Owen, and J. Stilgoe. 2013. 'Why Solar Radiation Management Geoengineering and Democracy Won't Mix'. *Environment and Planning A* 45, 2809–16.

Victor, D. 2008. 'On the Regulation of Geoengineering'. *Oxford Review of Economic Policy* 34(2): 322–36.

Victor, D.G., M.G. Morgan, F. Apt, and J. Steinbruner. 2009. 'The Geoengineering Option'. *Foreign Affairs* 88, 64–76.

Victor, D., G. Morgan, J. Apt, J. Steinbruger, and K. Rickie (2013). 'The Truth About Geoengineering' *Foreign Affairs*, 27 March. Available at: https://www.foreignaffairs.com/articles/global-commons/2013-03-27/truth-about-geoengineering.

Weitzman, M. 2011. 'Fat Tailed Uncertainty in the Economics of Catastrophic Climate Change'. *Oxford Review of Economic Policy* 5, 275.

## NOTES

1. I thank Rob Bellamy, Peter Healey, Steve Rayner and especially the volume's editors, Aaron Maltais and Catriona McKinnon, for very useful comments on an earlier version of this piece. The chapter was researched and written whilst I held a Leverhulme Early Career Fellowship, "Geoengineering and Global Justice" (ECF 2013-352), at the University of Warwick, and I am grateful to the Leverhulme Trust and the University for their support.

2. For example, in 2009, the Royal Society suggested that £10m of funding should be directed to research into geoengineering technologies. See Shepherd et al. (2009, xii).

3. This is complicated by the fact that it is difficult (some say imposssible) to reliably attribute specific changes to climate interventions.

4. For a comprehensive overview, see Rickels et al. (2011).

5. Critics of this view might attack the element of exceptionalism that SAI is 'incredibly' inexpensive (see discussion in Robert Pielke Jr [2010]; also Gordon McKerron [2014]). Barrett (2014) is more circumspect.

6. For an account of the rise of tipping points in climate discourse, see Russill and Nyssa (2009). Heyward and Rayner (2015) explore links between the 'climate tipping points' and 'climate emergency' rhetoric.

7. I shall not discuss the exceptionalist claim that SAI could be used unilaterally. For a discussion and an argument that the concern about unilateralism is largely grounded in 'unexamined policy assumptions' see Joshua Horton (2011).

8. It should be noted that the possible 'incompatibility with democracy' does not only apply to SAI. Ludvig Beckman (this volume) discusses whether citizens might be 'torn' between climate duties (conceived generally) and democratic duties.

9. See the project website: http://aerosols.ucsd.edu/E_PEACE.html.

10. See ICJ's ruling http://www.icj-cij.org/docket/files/148/18162.pdf.

11. The IPCC refrains from using language such as *climate catastrophe* and *climate emergency* and has in the main focused on linear climate changes.

12. Available at http://www.ce-conference.org/conference-videos. Transcribed by the author.

13. Rayner et al. offer five such values, embodied in the 'Oxford Principles' but stress that their list of values is intended to start a debate.

14. The test-bed was subsequently cancelled because of concerns about intellectual property.

## *Chapter Eight*

# Biomass Incineration

## *Scientifically and Ethically Indefensible*

Kristin S. Shrader-Frechette

Proponents of burning biomass to produce electricity have been doing a lot of 'greenwashing'. They try to claim biomass-energy technology is safe, sustainable and low carbon. Government often listens to their claims and the claims of biomass lobbyists. As a result, many governments subsidize biomass incineration. This chapter shows why such pro-biomass claims are likely false and such subsidies are misguided. It also shows that, once one takes account of all the health costs, environmental costs and economic costs, neither biomass incineration nor technologies like nuclear energy are defensible.

### THE DEBATE OVER BIOMASS INCINERATION

Prominent chemists proclaimed in 2012 that 'biomass has attracted a great deal of interest in recent years' (Serrano-Ruiz et al. 2012). The International Energy Agency likewise says that by 2020, biomass electricity generation will triple globally (International Energy Agency 2006). Even in the United States, biomass incineration is one of the largest non-hydro types of supposedly renewable energy, fulfilling government-mandated, renewable-energy credits (U.S. Energy Information Administration 2013; Booth 2012).

In the developing world, however, indoor incineration of biomass, including wood, dung and crop waste, poses a major health threat because at least 3 billion people, almost half the world's population, burn such dirty domestic biomass fuels. As a result, in the poorest countries 2 million people die prematurely each year from indoor air pollution, including 1 million deaths from chronic obstructive respiratory disease. Biomass-caused-particulate pollution is the major culprit, causing approximately 50 percent of all pneumonia deaths among children under the age of 5, mostly in the developing world (World Health Organization 2011).

Developed nations, however, say their biomass is a much cleaner energy source than what is used in typical indoor stoves in places like India (Booth 2012; Biomass Board 2013; Twisted Oak Corporation 2010; Shaddix 2011). They promote biomass incineration by claiming biomass is not only renewable but also fosters energy independence. As the U.S. Department of Energy (DOE) claims, promoters likewise say state-of-the-art, electricity-generating biomass facilities are clean and safe (U.S. Department of Energy 2012). As a result, countries like Canada, Denmark, England, Germany, Greece, Ireland, Italy, Japan, Portugal, Spain, Sweden, Turkey and the United States subsidize biomass/biofuels such as the reedy, grassy *Miscanthus giganteus* for electricity generation. They all provide biomass crop, biomass boiler construction and biomass renewable-energy subsidies for Miscanthus and other fuel crops (Sheehan et al. 2011; Department of Environment, Food and Rural Affairs 2007; Sustainable Energy Authority of Ireland 2012).

Using the example of Miscanthus giganteus, the next few sections show that state-of-the-art biomass facilities are neither safe nor clean mainly because, regardless of the type of biomass used, all such facilities involve combustion that produces particulate pollution that cannot be fully controlled and that has no safe dose. Given the health costs of biomass, those who pursue biomass electricity often do so for questionable ethical and economic reasons.

## STATE-OF-THE-ART BIOMASS FACILITIES

Miscanthus incineration for generating electricity deserves careful examination because Miscanthus is one of the three main biomass-crop alternatives and has some distinct advantages over the other two crops, corn and switchgrass. Miscanthus is dominant in Europe, and switchgrass is dominant in the United States, partly because it is the main biofuel crop that the U.S. DOE promotes and studies, although most current biomass energy is from wood.

This chapter investigates a proposed, state-of-the-art Miscanthus (rather than switchgrass or corn) facility for four main reasons. One reason is that the higher chlorine content of switchgrass—and therefore the potential both to slag the incinerator and to create dioxinsand furans when burned—make it somewhat less attractive than Miscanthus and corn (Ogden et al. 2010; El Nashaar et al. 2009). Second, the U.S. Department of Agriculture considers switchgrass to be a known and dangerous invasive species, whereas corn is not invasive and Miscanthus giganteus appears not to be invasive (Natural

Resources Conservation Services 2011). Third, corn—but not switchgrass or Miscanthus—is a food crop and developed nations recognize that they should not subsidize food crops for use as bioenergy (U.S. Department of Energy 2012). The consensus in the existing literature, as well as numerous statements by developing nations, indicates that using food crops for biomass will exacerbate global hunger and famine (Timilsina 2012). Fourth, for temperate climates, both corn and switchgrass have much lower productivity than does Miscanthus (Zhuang et al. 2013; Heaton et al. 2010.). Hence, if one wishes to investigate a leading biomass-incineration crop, arguably one should assess Miscanthus because it appears to have significant advantages that corn and switchgrass do not have. Unfortunately, however, incinerating Miscanthus and other biomass crops poses significant problems, perhaps the most significant of which is air pollution, a problem that all biomass-combustion facilities face.

## BIOMASS AIR POLLUTION

Because even the best of biomass crops use incineration in order to generate electricity, all such crops present serious air-pollution problems, even when they have the best available control technology for pollutants. Even in developed nations, biomass plants release carbon monoxide (CO), hazardous air pollutants such as mercury, nitrogen oxides (Shrader-Frechette and Preisser 2013), particulate matter (PM) and sulphur oxides (SOX). State-of-the-art, biomass-incineration facilities generally release pollutants whose harms are comparable to, or worse than, those from coal-fired plants, except that biomass facilities generally release less mercury and SOX, but more CO and PM. Yet PM has no safe dose (Pope et al. 2009). Moreover, as later paragraphs explain, biomass PM tends to be deadlier than coal PM, per kWhr of electricity, because biomass PM generally tends to include more smaller-size particles than does coal PM. Given this fact and given that PM causes most coal-incineration-related deaths, increased biomass PM of any kind is likely to outweigh most pollution improvements that biomass incineration might bring, relative to coal's mercury and SOX emissions (Booth 2012; Schneider 2000; Wiltsee 2000; Schneider 2004; Schneider and Banks 2010).

Biomass PM is likely far worse than that from coal because it contains more ultrafine particles, PMUF (<0.1 μm), whereas coal plant PM contains more fine particles, PMF (2.5–0.1 μm); by weight, PMUF is more dangerous than PMF because all PMUF is smaller in size than PMF and the smaller the particles, the more hazardous they are (mainly because they are more inflammatory) (National Institutes for Occupational Safety and Health 2013). Some typical types of PMUF—nanoparticles—are up to 65 times more hazardous than equal masses of PMF because PMUF harms are functions of surface area and numbers of particles, not mass or weight concentration (Sager and Castranova 2009). However, relative PMUF:PMF harms are functions of particle numbers/sizes/surfaces areas and both PMUF and PMF come in a variety of sizes. Consequently, the exact PMUF:PMF harms and fatalities are functions both of the specific sizes, numbers and surface areas of the precise particles in a given volume and of the specific facility, its pollution controls and the precise fuel burned. Consequently, only general-biomass: general-coal harm comparisons are possible in this chapter. Nevertheless, if leading U.S. government assessors are correct in stating that, on average, particulates (mostly PMF) from each U.S. coal plant kill 25 people/year (Schneider 2010), if some types of PMUF are up to 65 times more hazardous than some types of PMF, if all types of PMUF are more hazardous than the least-hazardous PMF (Sager and Castranova 2009) and if exposure routes for PMF and PMUF have roughly the same bodily penetration/exposure routes/velocity and so on, then an average, state-of-the-art, U.S. biomass plant kills far more than 25 people/year, as a coal plant does. Instead, because biomass incinerators produce more PMUF than do coal plants of the same size and because the smaller, lighter PMUF travels much farther and penetrates more deeply into the lungs, blood and other organs, biomass PM harm occurs over a much wider geographic range than does PMF harm from coal plants (National Institutes for Occupational Safety and Health 2013; Sager and Castranova 2009).

Although no nation has yet regulated PMUF, despite its far more serious health threat, the U.S. National Institutes for Occupational Safety and Health (NIOSH) admits that particle mass is not an adequate indicator of its hazards. NIOSH recently recommended occupational-only PMUF standards that are consistent with the previous suggested multiplier (65) for one specific type of PMF harms, in order to obtain PMUF harms. NIOSH says occupational-PMUF standards, for a 40-hour week, should be 8 times stricter than occupational-PMF standards (2.4 $mg/m^3$ for fine TiO2 and 0.3 $mg/m^3$ for ultrafine), to avoid causing greater than 1-in-1000, worker-cancer risks (National Institutes for Occupational Safety and Health 2011). If these proposed 40-hour/week NIOSH occupational standards are converted to standards for (four times longer) full-time exposure (168 hours/week), as they would need to be for a biomass-incineration facility, then they suggest that PMUF occupational regulations must be about (8)(4) or 32 times stricter, by mass, than PMF regulations, in order to protect against cancer risks greater than 1-in-1000 per year. However, these occupational standards are less protective than those for the general public, as regulations often protect the public against much lower risks, in the range of 1 in 100,000 to 1 in 1,000,000. Therefore, although NIOSH occupational recommendations suggest PMUF is likely about 32 times more dangerous, by mass, than is PMF, recommendations for public protection might need to be even stricter. Thus, the NIOSH occupational proposal may independently suggest that the earlier, experimentally-derived ratio of 1:65, for PMF and a typical type of PMUF, may be correct. After all, NIOSH warns that its 8-fold-stricter or 32-fold- stricter recommendations are (a) partial, purely occupational and (b) inapplicable to the more-sensitive public—who typically is protected against smaller-than 1-in-1000 risks. Given (b), the implicit, 32-times stricter NIOSH recommendation for PMUF is likely consistent with the 65-times riskier PMUF of a typical type, as compared to PMF.

Recognizing the death toll from coal plants, the previous considerations suggest that it hardly makes sense to implement a state-of-the-art biomass technology that could cause many more health problems than burning coal causes. How many deaths might biomass-plant UFPM cause each year?

Consider a very small, state-of-the-art Miscanthus-burning biomass facility, proposed for Jasper, Indiana. It would use 100,000 tons/year Miscanthus and release 0.03 pounds PMUP/million Btu, or 25 tons/year of mostly UFPM (Twisted Oak Corporation 2010; Shaddix 2011). If these 25 tons/year, Jasper-biomass-plant PMUF were PMF—which are much less hazardous than PMUF, as revealed earlier—government data show these PMF would cause the following additional, premature, avoidable harms/year (Schneider 2000; Schneider 2004): 1.6 deaths, 1.2 hospital admissions, 1.4 emergency-room visits, 2 heart attacks, 0.8 bronchitis cases, 29.2 asthma attacks and 167.7 lost work days.

As already noted, however, the ratio of PMUF:PMF harms is a function of particle surface area, number and so on. If so, one could obtain an approximate lower bound on the ratio of PMUF:PMF harms, given equal PM masses, when PMUF has the smallest surface area, least number of particles and largest-diameter size—namely, PMUF as close as possible to, but less than, 0.1 μm. Thus, for PMUF of just under 0.1 μm and PMF of 2.5 μm, PMUF:PMF harms are likely about 25:1 for equal masses of PMUF and PMF. What happens when one uses this multiplier of 25 to predict least-harmful PMUF effects, as a function of PMF? This multiplier shows that 25 tons of currently-unregulated PMUF from the small Jasper plant might cause roughly the following additional, premature, avoidable harms/year, across a wide geographic area—as far as the PM travels: 40 deaths, 30 hospital admissions, 35 emergency-room visits, 75 heart attacks, 20 bronchitis cases, 730 asthma attacks and 4,193 lost-work days (Shrader-Frechette 2013). Such figures suggest that, if one costs the annual deaths caused by PM from biomass incineration, as compared to those from coal PM—where PM is the deadliest pollutant released by both—it would be difficult to claim that biomass plants are economical in their health costs, relative to coal plants. This is in part because biomass appears deadlier than coal and yet government researchers already have shown that medical effects of fossil-fuel combustion, mostly from coal, include nearly 700,000 deaths/year globally and that climate effects (such as droughts, floods and hurricanes) of fossil fuels include 150,000 deaths/year globally (Shrader-Frechette 2011; Makhijani 2007; U.S. National Research Council 2009; Stern 2007; Working Group on Public Health and Fossil-Fuel Consumption 1997).

## POOR FARM ECONOMICS FOR BIOMASS

Another major biomass problem is that growing crops like Miscanthus is likely to be uneconomical, as compared to growing comparable crops such as corn and soy. Miscanthus requires far more water than they do, and it generates less profit for farmers than does corn or soy. For instance, much of the biomass-growing and temperate Midwest United States has had droughts and emergency water conservation in recent years—where Miscanthus presumably would be grown. There rainfall is about 20 percent less than the Miscanthus-required 30 inches of rain needed during a 7-month growing season (Heaton et al. 2012; Indiana Department of Natural Resources 2012; Singh and Kumar 2011; Purdue Extension, 2008). This means that because of heat-induced moisture losses and frequent negative values for precipitation (Charusombat and Niyogi 2011; NOAA National Climatic Data Center; North Dakota State University Department of Agriculture 1997; Mishra et al. 2009; Strezepek et al. 2010), Miscanthus likely would need to be irrigated even in temperate regions.

Indeed, Miscanthus yields are typically high only when Miscanthus is irrigated; yet Miscanthus irrigation is not cost effective (Heaton et al. 2012; Fargione et al. 2009; Clifton-Brown and Lewandoski 2000; Lewandowski et al. 2000). Thus, if farmers irrigate their Miscanthus, they likely will lose money—and if they do not irrigate, they could lose their Miscanthus altogether and thus again lose money (Kunycky and Shrader-Frechette 2013).

In fact, because of the higher cost of growing Miscanthus in relation to other crops, farmers typically refuse to grow it without price supports because doing so is not economical (Schill 2007; Khanna et al. 2008; Shrader-Frechette and Preisser 2013). Thus, not only could it be economically risky for farmers to grow Miscanthus crops, given drought problems, but also these economic and environmental problems might prevent farmers from growing enough Miscanthus to satisfy the demands of proposed plants, like that in Jasper. Both economic problems, in turn, could provide incentives for biomass-facility owners to cut their high fuel costs by using weaker pollution controls—especially given that PMUF is not yet regulated. If this weakening occurred, local deaths and injuries could rise even faster than already suggested.

Why is producing Miscanthus giganteus for biomass use likely more expensive for farmers than growing other crops, such as soybeans or corn?

Some of the reasons include (1) the difficulty establishing the Miscanthus plants; (2) the need for Miscanthus sellers to sell only 'plantlets', not seeds or rhizomes to farmers; (3) Miscanthus's very high water needs; (4) its drought intolerance, (5) its susceptibility to cold and (6) its inability to be planted and harvested with standard farm equipment—given that what is planted is not seeds—and that the harvested material comprises a dense, almost impenetrable, jungle. For instance, scientists have calculated the annualized operating costs of growing Miscanthus giganteus to be $988.88 per hectare and the yield to be 19.95 tons per hectare (Khanna et al. 2008). They have shown that corn and soybeans have annualized operating costs of $573.86 and $405.22 per hectare, respectively, with the yield of corn around 9.10 ton per hectare and the yield of soybeans around 3.14 ton per hectare (Khanna et al. 2008). If corn and soybean prices are $80.68 and $200.72 per ton, respectively (Khanna et al. 2008), the annual profit-per-hectare for corn is $160.33 and the annual profit-per-hectare for soybeans is $255.04. Based on these figures, in order for farmers to make the same profit-per-hectare on Miscanthus as they make for soybeans, the Miscanthus giganteus selling price must be about $61 per ton (Kunycky and Shrader-Frechette 2013). However, agricultural experts say the current (2012) price-per-ton of Miscanthus is only $38–40 (Heaton et al. 2012).

Of course, someone might suggest growing Miscanthus on non-crop land, land that is only marginal for growing crops, so that the supposed economic risks of growing Miscanthus might be lower (Miao and Khanna 2014). However, this argument makes no sense for at least three reasons. First, farmers are unlikely to purchase land in order to grow Miscanthus if they know they cannot also grow other crops on the same land, in case Miscanthus prices never improve. Second, the same factors that make land marginal for growing crops—such as lack of adequate water and lack of cold resistance—also make that land marginal for growing Miscanthus. Third, farmers are unlikely to grow a crop like Miscanthus given that it needs special harvesting equipment. Fourth, even if land is marginal for crops, it often is not marginal for developers, meaning that it could still be quite costly, even if growing Miscanthus on it is risky.

The preceding considerations and cost data mean that farmers may not even break even by planting Miscanthus. Indeed, based on earlier university and agricultural-extension data, they could lose about $200 per hectare if they grew Miscanthus instead of another crop. Hence, it is not surprising that, at least at present, most experts say that growing Miscanthus is not economical—unless the government subsidizes it (Ciciora 2011). Because farmers are more likely to choose to grow more profitable crops, the economic preference for growing food crops might prevent biomass plants, like that proposed for Jasper, Indiana, from having access to sufficient amounts of biomass to run the plant. Especially the data from the U.S. agricultural-extension services and agricultural universities suggest that only biomass-plant owners may profit from biomass incineration. Because owners typically create a limited-liability corporation for the biomass facility and because they enjoy massive taxpayer subsidies for their plants, they could benefit financially even if their biomass plant is not cost effective. Consumers, ratepayers and taxpayer subsidizers, however, might not benefit (Kunycky and Shrader-Frechette 2013).

## BIOMASS DROUGHT AND WATER-SHORTAGE THREATS TO OTHER WATER USERS

Growing nonsubsidized biomass crops like Miscanthus not only is likely uneconomical for farmers but also environmentally and economically dangerous for all citizens. One reason is that Miscanthus has a deeply penetrating (up to 2.5 meters, or 8.2 feet), dense, root mat (Werner 1995). This root mat would likely reduce groundwater availability in the already drought-prone Midwest (Boelcke et al. 1998). Moreover, the dense Miscanthus root mat is far more damaging than the root mat of almost any other plant. Miscanthus roots go beyond 8 feet, whereas 75 percent of the roots of switchgrass are in the top 0.3 meters (0.98 foot) of soil (McLaughlin et al. 1999). Even corn root mats are only about one-third as deep as Miscanthus; for corn, 90 percent of the roots are in the top 1 meter (3.28 feet) of soil (North Dakota State University Department of Agriculture 1997). Hence, Miscanthus is far more likely than corn to reduce groundwater availability for the local population (McIsaac et al. 2010).

As a result, growing Miscanthus could easily have damaging hydrological, and thus economic effects on the food and grain crops of local farmers. Moreover, it is questionable whether a biomass crop like Miscanthus ought to be allowed to deplete water resources that are needed for food crops. After all, Miscanthus has both a longer growing season and roughly a 12 percent lower intrinsic water-use efficiency than do food crops, like corn. Hence, all other things being equal, in a drought-prone region like the U.S. grain belt, one arguably ought to grow the crops that have higher water-use efficiency and shorter growing seasons (Dohleman and Long 2009). Otherwise, serious local economic losses could occur. For instance, the damaging 1988 drought that affected Indiana—home of the proposed Miscanthus incinerator discussed here—cost the country around $40 billion (Riebsame et al. 1991). The 2002 drought that affected Indiana costs the nation about $10 billion (Ross and Lott 2003). After the summer 2012 droughts in the Midwest, for instance, food prices were predicted to increase by 3–4 percent (Kunycky and Shrader-Frechette 2013; Volpe 2012).

Given the preceding four ways in which switchgrass is inferior to Miscanthus for biomass incineration, it also would make no sense to grow switchgrass instead. A key reason is that switchgrss has about double the water needs of Miscanthus—a fact that would exacerbate drought problems (Zhuang et al. 2013).

## WHY BIOMASS? THE LIKELY ROLE OF FLAWED ETHICS

In the face of the preceding problems with the dangerous health effects of biomass-burning facilities, their questionable economics and their threats to other land uses and to drought conditions, the obvious question is why so many areas are pursuing biomass incineration. Although it is impossible to reliably attribute intentions to people, one likely reason for the popularity of biomass—given the preceding problems—is government handouts and the fact that biomass incineration need not compete on the open market. That is, government subsidies for biomass provide a potential financial conflict of interest for biomass promoters. In addition, these powerful government subsidies also encourage promoters to downplay the health, economic and other risks of biomass, so as to site these facilities and to obtain government funding. As a result, citizens' rights to know about biomass health and economic harms may be jeopardized.

Regarding the taxpayer subsidizes that likely contribute to financial conflicts of interest, biomass-incineration plants, despite their questionable economics, need not operate on the market. They need not bring a profit because, at least in the United States, taxpayer-biomass subsidies are $3–5 billion/year (federal), in addition to $2–4 billion-per-plant/year (state). These enormous monies may help explain the 255 existing and 250 in-progress, U.S. biomass plants, despite the fact that they may not be cost-effective (Borsse et al. 2012). Such subsidies can induce local government officials and plant promoters to site the facilities, despite their health and economic problems.

This inducement and its resulting conflicts of interest seem clear, for instance, in the Jasper case. Jasper's two main alternatives to (a) the biomass facility are (b) using an old, dirty coal plant that, along with 250 others in the United States, otherwise would have to be closed because it fails to meet regulatory standards, or (c) using the same old, dirty coal plant but making a hefty outlay from the town of Jasper, required to make it comply with regulations. Given massive federal and state subsidies for biomass, option (a) would bring the town lease payments and $200 million locally over 30 years, according to the out-of-state, biomass-facility developer who has never done a biomass plant before (Twisted Oak Corporation 2010). Option (b) would leave the town with a liability, not an asset. Option (c) would be extraordinarily expensive for the town. Given these three options, financial considerations and financial conflicts of interest likely played a role in the Jasper decision to do (a), site the biomass plant and ignore its harmful public-health consequences.

Taxpayer subsidies for biomass and resulting financial conflicts of interest for local officials also likely jeopardize citizens' rights to know, as the Jasper case again suggests. Local Jasper politicians have agreed that, in return for a completely unspecified amount of lease payments from the proposed biomass plant, the town of Jasper would provide the biomass facility 'essential services', including double the water needed by the old coal facility, new electrical lines and other infrastructure (Twisted Oak Corporation 2010). However, local politicians have ensured that biomass lease-payment amounts were redacted from the proposed biomass-plant agreement that they presented to citizens (Twisted Oak Corporation 2010). Also redacted were lease terms, biomass-plant taxes, financing and town-borne costs of water/sewer/new electrical lines/contaminated materials (Twisted Oak Corporation 2010). Likewise redacted from the proposed biomass-plant agreement was all risk information from biomass-plant-partner, Mendel Bioenergy, about the Miscanthus giganteus seedlings that it would supply to local farmers—who would grow the crop for the proposed plant. Jasper local officials marked all redactions (in the proposed biomass agreement) as 'confidential materials'. Moreover, because this locally-polluting biomass facility, situated in the center of town, would sell its electricity on the open market and not to Jasper residents, Jasper residents would get no electricity benefits from the plant, despite bearing much of its economic and health costs (Twisted Oak Corporation 2010; Sheehan et al. 2011). The most plausible reason to explain why local officials would threaten citizens' rights to know such information is the financial rewards for which they hoped. They likely have financial conflicts of interest.

Indeed, threats to Jasper citizens' rights to know are so extreme that local residents do not even know precisely how much pollution the facility will release. For instance, the owners of the proposed Jasper facility promise to minimize its health/environmental impacts with baghouse filters, with biomass-boiler NOX/$CO_2$ best-available-control-technology limits, with 560 ppm CO limits and with 'voluntary' 0.03 lb./million (MM) Btu PM limits (Twisted Oak Corporation 2010; Shaddix 2011). However, their proposal is incomplete, in precisely the ways most likely to reveal flawed biomass-plant economics. The already-approved biomass proposal has no quantitative-human-health risk assessment (QRA), no cost-benefit analysis (CBA) and no ecological risk assessment (ERA). Therefore, no quantitative analysis of particulate and NOX harms, drought harms and so on, was ever performed and presented to citizens. Together, all four economic deficiencies suggest that, if local leaders have financial conflicts of interest and if communities are focused only on short-term economic benefits to a few people—e.g., facility lease payments—rather than long-term economic benefits to all, they might approve a likely dangerous and uneconomical biomass plant.

## OBJECTIONS

What objections might biomass proponents make to the preceding arguments that biomass incineration poses serious health threats, especially from ultrafine particulate pollution? What might they say in response to the claim that cultivating biomass crops like Miscanthus, as opposed to food crops like soybeans, is both uneconomical and environmentally destructive because of current drought conditions and Miscanthus's high water needs? Challengers might raise at least three questions: (1) Is the facility as dangerous as critics claim? (2) Can't Miscanthus and other biomass crops tolerate drought and therefore be more economical? (3) Don't some people say that, as long as Miscanthus crops are less than 25 percent of local land coverage, they will not threaten local groundwater and therefore food crops?

Regarding question (1), preceding sections of the chapter already presented data showing that biomass-incineration-plant owners themselves admitted, in official documents, that the biomass plant would release 0.03 pounds PMUF/million Btu, or 25 tons/year of mostly UFPM (Twisted Oak Corporation 2010; Shaddix 2011). Given earlier explanation of the fact that PM has no safe dose, that PMUF is the most dangerous PM, that PMUF is not regulated and that PMUF releases make virtually any biomass-incineration facility more dangerous than a coal facility, there are few grounds for the claim that biomass incineration is not harmful. Such a claim simply is contrary to the massive scientific evidence about the dangers of PM (Shrader-Frechette and Preisser 2013; Pope et al. 2009; Schneider 2000; Wiltsee 2000; Schneider 2004; Schneider 2010; National Institutes for Occupational Safety and Health 2013; Sager and Castranova 2009; National Institutes for Occupational Safety and Health 2011; Shrader-Frechette 2013).

Regarding question (2), Miscanthus promoters claim it can tolerate periods of drought and then return to a normal yield of 35.76 tons per hectare (Khanna et al. 2008) once normal climatic conditions occur the next year (Ivanic 2010). However, this 'return to normal' after a year of drought does not apply to the current climate-change situation. Climate change means that drought conditions likely will persist for many years at a time (Intergovernmental Panel on Climate Change 2007; YinPeng et al. 2009; Kundzewicz et al. 2008). Moreover, only established plants, not starting Miscanthus plants, could tolerate even one summer of drought. Besides, many scientists say Miscanthus is not drought-tolerant, even for a single season. They say it repeatedly has shown 'a lack of adaptation to drought' when put under water-stressed conditions (Clifton-Brown and Lewandowski 2000; Kunycky and Shrader-Frechette 2013; Lewandowski et al. 2000).

What about question (3), whether limiting Miscanthus crops to 25 percent of acreage would avoid drought-related economic, environmental and food harms? Miscanthus supporters say Miscanthus-caused decreases in groundwater availability occur when land cover near the biomass plant consists of over 25 percent Miscanthus giganteus (Vanloocke et al. 2010). In the case of the proposed Jasper, Indiana, plant, Miscanthus supporters claim that because the plant would require '10,000 acres . . . over a 35 to 50 mile radius from the plant' (Twisted Oak Corporation 2010), the Miscanthus land-coverage percentage would be less than 0.5 percent of the two counties within a 50 mile radius—and hence pose no drought threat to the local area.

Do the preceding responses to the obvious Micanthus-induced-drought risk succeed? They do not, for several reasons. First and most obviously, the answer—to the question whether or not 25-percent-Miscanthus coverage of some area is likely to harm groundwater supplies—depends on how one circumscribes the area and where the coverage is located. For instance, industry supporters of the proposed Jasper Miscanthus plant chose to use a circle, with a 50-mile radius, to describe the Miscanthus growing areas (Twisted Oak Corporation 2010). Yet if all 10,000 Miscanthus acres were within one contiguous portion of the 50-mile-radius circle, that portion obviously could have groundwater problems. Thus, without guarantees as to precisely where, in the circular area, Miscanthus might be grown, it could be concentrated in an area such that it would further induce drought. In other words, supporters of the 25-percent claim make the invalid assumption that, if Miscanthus will be grown in a given area, it will be uniformly distributed within that area. Obviously this assumption is false.

## ALTERNATIVES TO BIOMASS-GENERATED ELECTRICITY

If biomass electricity generation is not safer and healthier than coal, what is? Both Duke University and government surveys show that at least 80 percent of the public wants solar and wind development (Blackburn and Cunningham 2010; State of Massachusetts Energy and Environmental Affairs 2013; Shrader-Frechette 2011). Besides, in 2009 the chair of the US Federal Energy Regulatory Commission affirmed that, given smart-grid technology, new large coal, nuclear and biomass plants are not needed to meet any future electricity needs; all of these needs could be met cost-effectively with wind and other renewables (Straub and Behr 2009). Nuclear plants especially are not needed because, at least in the United States, they have received 96 percent of all subsidies for nuclear, solar and wind energy and yet they are currently more expensive than solar, far more expensive than wind and likely to become even more so. In addition, even without accidents, all nuclear plants increase nearby cancer rates because of normal radiation releases, and they are likely to cause harm to distant generations given no viable solution for long-term waste disposal. Besides, if nuclear plants were really safe, virtually all nations would not have given the nuclear industry a liability limit to protect it against bankruptcy following an accident (Shrader-Frechette 2011).

Investors (Deutsche Bank Group 2011), credit-rating agencies (Moody's Analytics 2009) and business magazines (Fahey 2010; Morton 2012) say something similar about the problems with nuclear and biomass and the promise of wind and solar. As a classic Princeton University article, published in *Science* and updated in a 2011 issue of *Bulletin of the Atomic Scientists*, says, only 7–9 renewable or efficiency technologies, 'already deployed at an industrial scale', could easily and quickly halt global warming (Pacala and Socolow 2004; Socolow 2011). Duke University economists showed that since 2010, solar-photovoltaic electricity costs have been at least 4–6 cents cheaper, per kWhr, than electricity from new nuclear plants which is now 20–21 cents/kWhr and increasing (Blackburn and Cunningham 2010). Moreover, wind is even cheaper and safer than solar photovoltaic. Already by 2003, the International Energy Agency said that wind energy cost 2–3 cents per kWhr and was dropping, making wind several times cheaper than nuclear power (International Energy Agency 2003; Smith 2006; Shrader-Frechette 2011). Indeed, by 2008, in many places in the United States, wind costs were below one cent per kWhr (International Energy Agency 2009; Patel 2006). These low wind prices are why, for the last 5 years, installed wind has been growing faster, globally, than any other power source (Renewable Energy Policy Network for the 21st Century 2008; Mouawad 2010).

Wind also has almost no adverse environmental impacts. The classic 2007 U.S. National Academy of Sciences report on environmental effects of wind energy denies that wind turbines are a threat to flora and fauna, including birds. Indeed, the report explicitly states that wind energy has the fewest 'adverse environmental impacts' of any 'sources of energy' used for 'electricity generation' (National Research Council 2007). By 2012, wind and solar-PV together supplied 20 percent of Germany's electricity (Burger 2013), wind supplied more than 30 percent of Denmark's electricity and in 2020 will supply 50 percent (Danish Wind Energy Association 2013). By 2012, wind also supplied 25 percent of Iowa's electricity, 24 percent in South Dakota and so on—although these are not the largest-wind-resource U.S. states (Iowa Wind Energy Association 2010; Roney 2013). Given such economic data, in 2012 wind accounted for more new U.S. power capacity than coal, natural gas, nuclear, 'or anything else. It made up 42 percent of new power capacity additions in the United States' in 2012 (American Wind Energy Association 2013).

## CONCLUSIONS

Obviously there is no space here for a complete solar-wind-conservation-efficiency analysis as an alternative to biomass-generated and other dirty electricity, such as coal and nuclear. Yet, such an analysis is available elsewhere (Shrader-Frechette 2011; Pacala and Socolow 2004; Socolow 2012; Makhijani 2007). In conclusion, given the massive air-pollution, economics and drought problems associated with growing and burning biomass crops like Miscanthus and given the viable alternatives to biomass incineration, the future belongs to clean, renewable energy, not to dirty energy—not to anything that requires incineration and thus pollutes the air. Even if dirty biomass energy is locally produced and sustainable, it is not desirable, given cleaner and cheaper alternatives.

## ACKNOWLEDGEMENTS

Thanks to University of Notre Dame biologists Brianna Kunycky and Whitney Preisser for some of the research on biomass energy technology.

## CONFLICT OF INTEREST

The author declares no conflict of interest.

## REFERENCES

American Wind Energy Association (AWEA). 2013. *U.S. Wind Industry Fourth Quarter 2012 Market Reports*. Washington, DC: AWEA. http://www.awea.org/learnabout/publications/reports/upload/AWEA-
Fourth-Quarter-Wind-Energy-Industry-Market-Report_Executive-Summary-4.pdf.

Biomass Board. 2013. *Biomass Research and Development: Advancing Bioenergy Technologies*. Washington, DC: US Department of Energy.

Blackburn, J. O., and S. Cunningham. 2010. *Solar and Nuclear Costs, The Historic Crossover: Solar Energy Is Now the Better Buy*. Durham, NC: NCWARN.

Boelcke, B., S. Buech, and S. Zacharias. 1998. 'Effects of Miscanthus Cultivation on Soil Fertility and the Soil Water Reservoir'. In *Biomass for Energy and the Environment*. 911–14. Wurzburg, Germany: Rimpar.

Booth, M. S. 2012. *Biomass Air Pollution*. Springfield, MA: Partnership for Policy Integrity. http://www.pfpi.net/air-pollution-2.

Borsse, N., A. Dufour, X. Meng, Q. Sun, and A. Ragauskas. 2012. 'Miscanthus: A Fast-Growing Crop for Biofuels and Chemicals Production'. *Biofuels, Bioproducts and Biorefining* 6 (5): 580–98.

Burger, B. 2013. *Electricity Production from Solar and Wind in Germany in 2013.* Freiberg, Germany: Fraunhofer Institute. http://www.ise.fraunhofer.de/en/downloads-englisch/pdf-files-englisch/news/electricity-production-from-solar-and-wind-in-germany-in-2013.pdf

Charusombat, U., and D. Niyogi. 2011. 'A Hydroclimatological Assessment of Regional Drought Vulnerability: A Case Study of Indiana Droughts'. *Earth Interactions* 15: 1–65.

Ciciora, P. 2011. *A Billion Tons of Biomass a Viable Goal, But at High Price, New Research Shows.* Accessed November 5, 2012. http://www.renewableenergyworld.com/rea/news/article/2011/03/a-billion-tons-of-biomass-a-viable-goal-but-at-high-price-new-research-shows.

Clifton-Brown, J., and I. Lewandowski. 2000. 'Water Use Efficiency and Biomass Partitioning of Three Different Miscanthus Genotypes with Limited and Unlimited Water Supply'. *Annals of Botany* 86: 191–200.

Department of Environment, Food and Rural Affairs (DEFRA). 2007. *Planting and Growing Miscanthus*. Accessed December 12, 2012. www.defra.gov.uk/erdp.

Deutsche Bank Group. 2011. *The 2011 Inflection Point for Energymarkets: Health, Safety, Security and the Environment.* New York, NY: DB Climate Change Advisors.

Dohleman, F., and S. Long. 2009. 'More Productive than Maize in the Midwest: How Does Miscanthus Do It?' *Plant Physiology* 150: 2104–115.

El Nashaar, H. M., G. M. Banowetz, S. M. Griffith, M. D. Casler, and K. P. Vogel. 2009. 'Genotypic Variability in Mineral Composition of Switchgrass'. *Bioresource Technology* 100: 1809–14.

Fahey, J. 2010. 'Southern Co'.s Nuclear Game Plan'. *Forbes*, September 9. http://www.forbes.com/forbes/ 2010/0927/energy-technology-nuclear-power-southern-company.html

Fargione, J. E., T. R. Cooper, D. J. Flaspohler, J. Hill, C. Lehman, D. Tilman, T. McCoy, S. McLeod, E. J. Nelson, and K. S. Oberhauser. 2009. 'Bioenergy and Wildlife: Threats and Opportunities for Grassland Conservation'. *BioScience* 59: 767–77.

Heaton, E., N. Boersma, J. Caveny, T. Voigt, and F. Dohleman. 2012. *Miscanthus (Miscanthus x giganteus) for Biofuel Production. Purdue University Purdue Extension.* Purdue, IN: Purdue University. http://www.extension.org/pages/26625/miscanthus-miscanthus-x-giganteus-for-biofuel-production.

Heaton, E.A., F. G. Dohleman, A. F. Miguez, J. A. Juvik, V. Lozovaya, J. Widholm, O. A. Zabotina, G. F. Mcisaac, M. B. David, T. B. Voigt, N. N. Boersma, and S. P. Long. 2010. 'Miscanthus: A Promising Biomass Crop'. *Advances in Botanical Research* 56: 75–137.

Indiana Department of Natural Resources. 2012. *Monthly Water Resource Survey*. Accessed September 13, 2012. http://www.in.gov/dnr/water/4858.htm.

Intergovernmental Panel on Climate Change (IPCC). 2007. *Fourth Assessment Report AR4*. Geneva, Switzerland: IPCC. http://www.ipcc.ch/publications_and_data/publications_and_data_reports.shtml#.UGc1ztBYuWM.

International Energy Agency (IEA). 2003. *Renewables for Power Generation: Status and Prospects*. Paris, France: IEA.

———. 2006. *Renewables in Global Energy Supply*. Paris, France: Organization for Economic Cooperation and Development.

———. 2009. *Technology Roadmap: Wind Energy*. Paris, France: IEA. http://www.iea.org/papers/2009/Wind_Roadmap.pdf.

Iowa Wind Energy Association. 2010. *Wind Power Facts*. Accessed July 2, 2013. http://www.iowawindenergy.org/whywind.php.

Ivanic, R. 2010. *Miscanthus Establishment Protocol, Fertilizer Application and Water Requirements*. Millen, GA: Twisted Oak Corporation. http://www.jasperindiana.gov/images/contentimages/10acy4e4b.pdf.

Khanna, M., B. Dhungana, and J. Clifton-Brown. 2008. 'Costs of Producing Miscanthus and Switchgrass for Bioenergy in Illinois'. *Biomass and Bioenergy* 32: 482–83.

Kundzewicz, Z., L. Mata, N. Arnell, P. Döll, B. Jimenez, K. Miller, T. Oki, Z. Sen, and I. Shiklomanov. 2008. 'The Implications of Projected Climate Change for Freshwater Resources and Their Management'. *Hydrological Sciences Journal* 53: 3–10.

Kunycky, B. N., and K. S. Shrader-Frechette. 2013. 'Lessons on Drought and Pollution from the Forgotten Three Billion'. *Global Health Perspectives* 1: 55–62.

Lewandowski, I., J. Clifton-Brown, J. Scurlock, and W. Huisman. 2000. 'Miscanthus: European Experience with a Novel Energy Crop'. *Biomass and Bioenergy* 19: 209–27.

Makhijani, A. 2007. *Carbon-Free and Nuclear-Free*. Takoma Park, MD: Institute for Energy and Environmental Research.

McIsaac, G., M. David, and C. Mitchell. 2010. 'Miscanthus and Switchgrass Production in Central Illinois: Impacts on Hydrology and Inorganic Nitrogen Leaching'. *Environmental Quality* 39: 1790–99.

McLaughlin, S., J. Bouton, D. Bransby, B. Conger, W. Ocumpaugh, D. Parrish, C. Taliaferro, K. Vogel, and S. Wullschleger. 'Developing Switchgrass as a Bioenergy Crop'. In *Perspectives on New Crops and New Uses*, edited by J. Janick, 282–99. Alexandria, VA: ASHS Press.

Miao, R., and M. Khanna. 2014. 'Are Bioenergy Crops Riskier than Corn? Implications for Biomass Price'. *Choices* 1st Quarter. http://www.choicesmagazine.org/choices-magazine/theme-articles/economic-and-policy-analysis-of-advanced-biofuels/are-bioenergy-crops-riskier-than-corn-implications-for-biomass-price.

Mishra, V., K. Cherkauer, and S. Shukla. 2009. 'Assessment of Drought Due to Historic Climate Variability and Projected Future Climate Change in the Midwestern United States'. *Journal of Hydrometeorology* 2: 46–68.

Moody's Analytics. 2009. *Infrastructure Special Comment: Right-Way Hedging for Power Companies*. Accessed June 5, 2012. http://www.moodys.com/moodys/cust/research/MDCdocs/24/2007400000621491.pdf?search= 6&searchQuery=hess&click=1.

Morton, O. 2012. 'The Dream That Failed'. *The Economist*, March 10. www.economist.com/node/21549098.

Mouawad, J. 2010. 'Wind Power Grows 39 Percent for the Year'. *The New York Times*, January 26.

National Institutes for Occupational Safety and Health. 2011. *Occupational Exposure to Titanium Dioxide.* Atlanta, GA: U.S. Centers for Disease Control.

———. 2013. *Occupational Exposure to Carbon Nanotubes and Nanofibers*. Atlanta, GA: U.S. Centers for Disease Control.

National Research Council. 2007. *Environmental Impacts of Wind-Energy Projects*. Washington, DC: National Academy Press.

Natural Resources Conservation Services. 2011. *Plant Fact Sheet: Switchgrass.* Accessed July 27, 2013. http://plants.usda.gov/factsheet/pdf/fs_pavi2.pdf.

NOAA National Climatic Data Center. 2012. *State of the Climate: Drought*. Accessed September 24, 2012. http://www.ncdc.noaa.gov/sotc/drought/.

North Dakota State University Department of Agriculture. 1997. *Corn Production Guide*. Accessed November 2, 2012. http://www.ag.ndsu.edu/pubs/plantsci/rowcrops/a1130-8.htm.

Ogden, C. A., K. E. Ileleji, K. D. Johnson, and Q. Wang. 2010. 'In-field Direct Combustion Fuel Property Changes of Switchgrass Harvested from Summer to Fall'. *Fuel Processing Technology* 91: 266–71.

Pacala, S., and R. Socolow. 2004. 'Stabilization Wedges: Solving the Climate Problem for the Next 50 Years with Current Technologies'. *Science* 305: 968–72.

Patel, M. R. 2006. *Wind and Solar Power Systems—Design, Analysis and Operation*, 2nd ed. Boca Raton, FL: Taylor and Francis Group.

Pope, C.A., R. T. Burnett, D. Krewski, M. Jerrett, Y. Shi, E. E. Calle, and M. J. Thun. 2009. 'Cardiovascular Mortality and Exposure to Airborne Fine Particulate Matter and Cigarette Smoke: Shape of the Exposure-Response Relationship'. *Circulation* 120: 941–48.

Purdue Extension. 2008. *Corn Water Requirements*. Accessed November 2, 2012. http://www.extension.org/pages/14080/corn-water-requirements.

Renewable Energy Policy Network for the 21st Century (REN21). 2008. *Renewables 2007: Global Status Report*. Paris, France: REN21.

Riebsame, W., S. Changnon, and T. Karl. 1991. *Drought and Natural Resources Management in the United States: Impacts and Implications of the 1987-89 Drought*. Boulder, CO: Westview Press.

Roney, J. M. 2013. 'Iowa and South Dakota Approach 25 Percent Electricity from Wind in 2012: Unprecedented Contribution of Wind Power in U.S. Midwest'. Washington, DC: Earth Policy Institute. http://www.earth-policy.org/data_highlights/2013/highlights37.

Ross, T., and N. Lott. 2003. *National Climatic Data Center Technological Report*. Asheville, NC: National Oceanic and Atmospheric Administration.

Sager, T. M., and V. Castranova. 2009. 'Surface Area of Particle Administered Versus Mass in Determining the Pulmonary Toxicity of Ultrafine and Fine Carbon Black: Comparison to Ultrafine Titanium Dioxide'. *Particle and Fibre Toxicology* 6: 15–21.

Schill, S. R. 2007. 'Miscanthus versus Switchgrass'. *Ethanol Producer Magazine*, October 3. 2007. http://www.ethanolproducer.com/articles/3334/miscanthus-versus-switchgrass.

Schneider, C. G. 2000. *Death, Disease and Dirty Power*. Boston, MA: Abt Associates.

Schneider, C. G. 2004. *Dirty Air, Dirty Power*. Boston, MA: Clean Air Task Force/Abt Associates.

Schneider, C. G., and J. Banks. 2010. *The Toll from Coal: An Updated Assessment of Death and Disease from America's Dirtiest Energy Source*. Washington, DC: Abt Associates.

Serrano-Ruiz, J. C., R. Luque, J. M. Campelo, and A. A. Romero. 2012. 'Continuous-flow Processes in Heterogeneously Catalyzed Transformations of Biomass Derivatives into Fuels and Chemicals'. *Challenges* (3): 115–18.

Shaddix, C. S. 2011. *Review of Twisted Oak Proposal and Associated Literature.* Jasper, IN: Jasper Utilities Board.

Sheehan, M., S. Chirillo, J. Schlossberg, W. Sammons, and W. Leonard. 2011. *Biomass Electricity: Clean Energy Subsidies for a Dirty Industry*. Cambridge, Massachusetts: Biomass Accountability Project.

Shrader-Frechette, K. S. 2011. *What Will Work: Fighting Climate Change with Renewable Energy, Not Nuclear Power*. New York, NY: Oxford University Press

———. 2013. 'Biomass and Effects of Airborne Ultrafine Particulates: Lessons about State Variables in Ecology'. *Biological Theory* 8: 44–48.

Shrader-Frechette, K. S., and W. Preisser. 2013. 'Renewable Technologies and Environmental Injustice: Subsidizing Bioenergy, Promoting Inequity'. *Environmental Justice* 6 (3): 88–93.

Singh, S., and A. Kumar. 2011. 'Development of Water Requirement Factors for Biomass Conversion Pathway'. *Bioresource Technology* 102: 1316–28.

Smith, B. 2006. *Insurmountable Risks: The Dangers of Using Nuclear Power to Combat Global Climate Change*. Takoma Park, MD: IEER Press.

Socolow, R. 2012. 'Wedges Reaffirmed'. *The Bulletin of Atomic Scientist*, September 27. http://www.the bulletin.org/web-edition/features/wedges-reaffirmed.

State of Massachusetts Energy and Environmental Affairs. 2013. *Wind Energy Facts*. Accessed June 26, 2013. http://www.mass.gov/eea/energy-utilities-clean-tech/renewable-energy/wind/wind-energy-facts.htm.

Stern, N. 2007. *The Economics of Climate Change*. Cambridge: Cambridge University Press.

Straub, N., and P. Behr. 2009. 'Will the U.S. Ever Need to Build Another Coal or Nuclear Power Plant?' *Scientific American*, April 22. http://www.scientificamerican.com/article.cfm?id=will-the-us-need-new-coal.

Strzepek, K., G. Yohe, J. Neumann, and B. Boehlert. 2010. 'Characterizing Changes in Drought Risk for the United States from Climate Change'. *Environmental Research Letters* 5: 1–9.

Sustainable Energy Authority of Ireland (SEA). 2012. *Fact Sheet, Miscanthus.* Dublin, Ireland: Renewable Energy Information Office, SEA.

Timilsina, G. R. 2012. 'BioFuels'. In *Plant Sciences Reviews*, edited by D. Hemming, 171–78. Boston, MA: CABI.

Twisted Oak Corporation (TOC). 2010. *Site Lease and Repowering Proposal: Submitted to City of Jasper Municipal Utility Department*. Sandy Springs, GA: TOC. http://duboiscountyherald.com/c/vo_file_private/03/a5/71d685147128ef4e5e8aaf7c29f8.pdf.

U.S. Department of Energy (DOE). 2012. *Biomass Program*; DOE. Accessed January 4, 2013. http://www1.eere.energy.gov/biomass/.

U.S. Energy Information Administration (EIA). 2013. *Biomass for Electricity Generation.* Washington, DC: U.S. Department of Energy. http://www.eia.gov/oiaf/analysispaper/biomass/.

U.S. National Research Council. 2009. *Hidden Costs of Energy.* Washington, DC: National Academy Press, 2009.

Vanloocke, A., C. Bernacchi, and T. Twines. 2010. 'The Impacts of *Miscanthus x giganteus* Production on the Midwest United States Hydrologic Cycle'. *Global Change Biology Bioenergy* 2: 180–91.

Volpe, R. 2012. *Consumer Price Index Forecast.* Washington, DC: United States Department of Agriculture Economic Research Service. http://www.ers.usda.gov/data-products/food-price-outlook.aspx.

Werner, I. 1995. 'Umweltaspekte im Miscanthusanbau — wurzeluntersuchungen, phytosanitäre untersuchungen and untersaaten'. In *Symposium Miscanthus — Biomassebereitstellung, Energetische und Stoffliche Nutzung*. 87-101. Münster, Germany: Landwirtschaftsverlag.

Wiltsee, G. 2000. *Lessons Learned from Existing Biomass Power Plants*. Valencia, CA: National Renewable Energy Laboratory.

Working Group on Public Health and Fossil-Fuel Combustion. 1997. 'Short-term Improvements in Public Health from Global-Climate Policies on Fossil-Fuel Combustion' *The Lancet*. 350 (9088): 1341–49.

World Health Organization (WHO). 2011. *Indoor Air Pollution and Health*. Geneva, Switzerland: WHO. http://www.who.int/mediacentre/factsheets/fs292/en/index.html.

YinPeng, L., Y. Wei, W. Meng, and Y. XiaoDong. 2009. 'Climate Change and Drought: A Risk Assessment of Crop-Yield Impacts'. *Climate Research* 39: 31–46.

Zhuang, Q., Z. Qin, and M. Chen. 2013. 'Biofuel, Land and Water: Maize, Switchgrass or Miscanthus'? qzhuang@purdue.eduqin9@purdue.edu[1] *Department of Earth, Atmospheric and Planetary Sciences, Purdue University, West Lafayette, IN 47907, USA* [2] *Department of Agronomy, Purdue University, West Lafayette, IN 47907, USA* [3] *These authors contributed equally to this work. Environmental Research Letters* 8. http://iopscience.iop.org/1748-9326/8/1/015020.

# Index

# Notes on the Contributors

**Ludvig Beckman** is Professor at the Department of Political Science, Stockholm University (Sweden). His recent books include *The Territories of Citizenship* (2012) (edited with Eva Erman) and *The Frontiers of Democracy: The Right to Vote and Its Limits* (2009). He has published widely on democratic theory, intergenerational justice, migration, children's rights, privacy and bioethics. He is currently the Vice Head of Department.

**Megan Blomfield** is a Lecturer in value theory in the Department of Philosophy, University of Bristol. Her major area of research is global justice and the environment, with a focus on the problem of climate change; rights to natural resources; and how different factors—such as scientific uncertainty—should influence collective action to address environmental problems. Her chapter in this volume was written during a post-doctoral fellowship at Stanford University's Center for Ethics in Society.

**Clare Heyward** is a Leverhulme Early Career Fellow at the University of Warwick. Her current project is *Global Justice and Geoengineering*. Clare completed her DPhil at the University of Oxford. Her thesis argued that climate change could be regarded as a form of cultural injustice and explored how taking the cultural dimension of climate justice seriously might affect societal responses to climate change. Before joining the University of Warwick, she was James Martin Research Fellow on the Oxford Geoengineering Programme. Clare is interested in issues of global distributive justice and intergenerational justice, especially those connected to climate change.

**Jonathan Kuyper** is a Senior Postdoctoral Research Fellow at Stockholm University. He completed his PhD at The Australian National University in

2012 and has held visiting researcher positions at the European University Institute, the University of Canberra and Princeton University. His research centres on normative issues in world politics with a specific focus on intellectual property rights and climate governance. His work has appeared, or is forthcoming, in the *European Journal of International Relations*, *Political Studies*, *Critical Policy Studies*, *Global Constitutionalism*, *Swiss Political Science Review*, *Critical Review*, and *Ethics and Global Politics*.

**Aaron Maltais** is a Postdoctoral Fellow at the Department of Political Science at Stockholm University. He specialises in contemporary political theory and his main research interests are in the politics and ethics of climate change, theories of global justice and theories of political obligation. Aaron's work on climate change addresses questions of fairness and justice that arise as societies attempt to make climate governance more effective and as they transition to low-carbon energy. His work on climate change has been published in *Environmental Politics, Environmental Values,* and *Political Studies.* Aaron also does work on political obligations and the moral legitimacy of political institutions. He has recently published a theory of political obligations in *Legal Theory* and is now working on the implications of his approach for the legitimacy of immigration controls.

**Catriona McKinnon** is Professor of Political Theory in the Department of Politics and International Relations, University of Reading. She is the author of *Climate Change and Future Justice* (2012) and *Toleration: A Critical Introduction* (2006) and co-editor (with Gideon Calder) of *Climate Change and Liberal Priorities* (2012). She is currently finishing a book on climate change and international criminal law. She is the Director of the Leverhulme Programme in Climate Justice at the University of Reading.

**Kristin Shrader-Frechette** has degrees in mathematics and philosophy of science and has held three U.S. National Science Foundation (NSF)–funded post-doctoral fellowships, in biology, economics and hydrogeology. O'Neill Professor at University of Notre Dame, she has held senior professorships at the University of California and the University of Florida. Her research, funded continuously for 28 years by the U.S. NSF, addresses flawed models in biology/hydrogeology; default rules under mathematical/scientific uncertainty; quantitative risk analysis; and science and values/ethics. Her work has been translated into 13 languages and includes 15 books such as *Tainted* (how flawed scientific methods influence science policy); *What Will Work: Fighting Climate Change with Renewable Energy, Not Nuclear Power; Taking Action, Saving Lives*; *Ethics of Scientific Research*; *Method in Ecology; Risk Analysis and Scientific Method, and Risk and Rationality.* Her 400+ journal articles appear in *Biological Theory, Bioscience, Philosophy of*

*Science, Quarterly Review of Biology, Bulletin of the Atomic Scientists, Energy Policy, Modern Energy Review, Risk Analysis, Ethics, Public Affairs Quarterly*, and *Science* (3 pieces). She has served on many international and U.S. Department of Energy, Environmental Protection Agency, and National Academy of Sciences boards/committees. Her pro-bono scientific/ethics work, to protect poor/minority communities from pollution-caused environmental injustice, have won her many awards, including the World Technology Association's Ethics Prize and Tufts University's Global Citizenship Award.

**Patrick Taylor Smith** received his PhD in Philosophy from the University of Washington, Seattle, where he wrote a dissertation on the legitimacy of global governance institutions. He has been a Postdoctoral Fellow at Stanford University's Center for Ethics in Society, and is now Assistant Professor of Global Studies at National University Singapore. He works in global and intergenerational justice, with more specific interests in climate change, geoengineering, and global governance. He has published articles in journals such as *Philosophy and Public Issues, Stanford Encyclopedia of Philosophy, Ethical Theory and Moral Practice, Notre Dame Philosophical Reviews,* and *Ethics, Policy, and Environment* as well as book chapters on solar radiation management and cyberwarfare.

**Steve Vanderheiden** is Associate Professor of political science and environmental studies at the University of Colorado at Boulder in the United States and Professorial Fellow with the Centre for Applied Philosophy and Public Ethics (CAPPE) at Charles Sturt University in Canberra, Australia. Trained in political science and philosophy at the University of Wisconsin-Madison (PhD 2001), his areas of research expertise include normative political theory and environmental politics, with a concentration in justice issues in global climate governance. He has published numerous articles and book chapters on topics ranging from the environmental thought of Rousseau to the framing of contemporary inequalities in access to ecological services as issues of environmental justice. His book *Atmospheric Justice: A Political Theory of Climate Change* (Oxford University Press, 2008) won the 2009 Harold and Margaret Sprout award, given by the International Studies Association to the best book of the year in environmental politics. His current projects include an empirical study of the climate change adaptation finance mechanisms being developed under the auspices of the United Nations Framework Convention on Climate Change, an analysis of informational governance in the context of climate change policy, and a book-length treatment of problems associated with individual responsibility for global environmental problems like climate change.